Guide to Services Selection and Integration in Low Rise Buildings

Guide to Services Selection and Integration in Low Rise Buildings

Jim Alder

Mitchell, London

First published 1983
First paperback edition 1987

Typeset by Tek-Art Ltd, Kent
and printed in Great Britain by
R.J. Acford, Chichester, Sussex.
Published by The Mitchell Publishing Company Limited
4 Fitzhardinge Street, London W1H 0AH
A subsidiary of B.T. Batsford Limited

ISBN 0 7134 5746 5

D
696.1
ALD

Contents

Preface

The object of this book is to assist architectural and building students, and those already engaged in design of domestic, commercial and small industrial buildings, who have only a very basic knowledge of the practicalities of service selection and the integration of them into the structure. It attempts to point out some of the many problems which can be reasonably anticipated at the initial design stage regarding the service itself and the space which is necessary for it, together with the integration of it with the many structural units, materials and finishes. Also mention is made of some of the unsatisfactory results which may arise due to systems or items having been specified without sufficient consideration being given initially.

This text is intended to point out the advantages, disadvantages and limitations of known service systems when related to the various environments, and so help guide all those concerned into making reasonable selections for the type of structure proposed, taking into consideration the site and the building use envisaged. The annotated sketches are not to scale but intended to act as a useful refresher as well as amplifying the particular point or topic under consideration.

The purpose is also to assist those concerned in ensuring that it is possible and practical to install the systems proposed with the type of construction, materials and finishes specified. It attempts to make the building design team aware of the service maintenance needed and so alert them early on about the necessary provisions, for example the need of adequate space for the replacement of plant, which should be given attention at the design stage. It is hoped that all those concerned with construction and services will not only consider their own work but also show some thought and understanding for other people's jobs, the client who has to live with the completed product, and last but not least the man who has to carry out the routine maintenance later. Faults may arise but the results are often made worse due to a lack of appreciation of other craftsmen's work and basic needs, and in many cases a failure to apply pure common sense.

The idea for this book arose from observations made on visits to buildings both new and old, and from answering students' questions, and offering constructive criticism, while they were engaged on practical projects.

Appreciating a need, I have therefore compiled this text for those who are not service engineers but who are involved to a limited extent with building services, to make them aware of the basic problems which can arise, and to offer some guidance on steps which may be taken to avoid their occurrence or reduce their effects. I hope it will also help to reduce trouble arising from building services, make maintenance easier, and avoid the frustration and unnecessary expense which so often accompanies it. If it does, then I shall feel that the effort has been well worth while.

Jim Alder
Brighton 1983

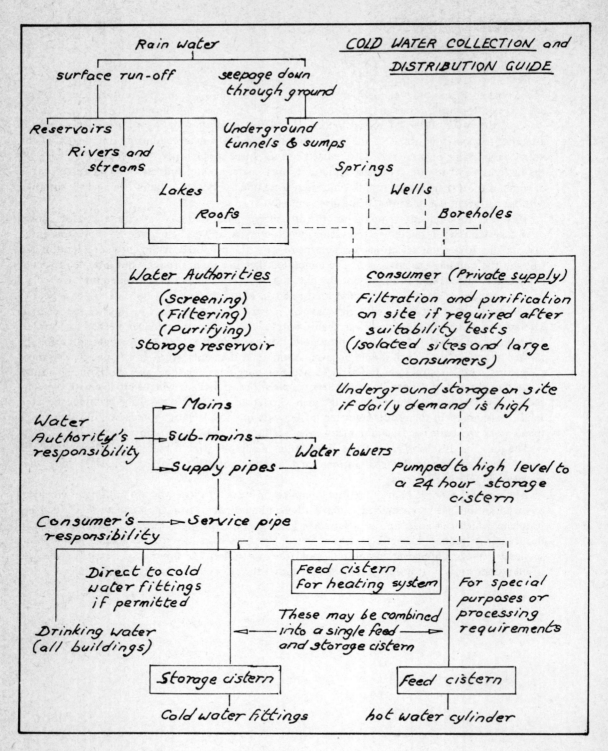

Rain water

COLD WATER COLLECTION and DISTRIBUTION GUIDE

surface run-off   seepage down through ground

Reservoirs   Underground tunnels & sumps
    Rivers and streams   Springs
            Lakes   Wells
                Roofs   Boreholes

Water Authorities
(Screening)
(Filtering)
(Purifying)
storage reservoir

consumer (Private supply)
Filtration and purification on site if required after suitability tests (isolated sites and large consumers)

Underground storage on site if daily demand is high

Water Authority's responsibility
→ Mains
→ Sub-mains
→ Supply pipes   Water towers

Pumped to high level to a 24 hour storage cistern

Consumer's responsibility ──▷ Service pipe

Direct to cold water fittings if permitted

Feed cistern for heating system

For special purposes or processing requirements

These may be combined into a single feed and storage cistern

Drinking water (all buildings)

Storage cistern

Feed cistern

Cold water fittings

hot water cylinder

8

1 Cold water supply

Water sources may be rivers, lakes, natural reservoirs in the hills or artificially constructed ones below ground but located at the high spots in the local terrain and supplied with water pumped up from collection points in tunnels cut in the chalk. Springs, wells and boreholes are also used for isolated buildings and special needs in industry.

The water is generally gravity fed direct to the buildings through a system of pipes and is suitable for drinking having already been screened to take out any floating debris, filtered to remove any impurities remaining in suspension and finally chemically treated to kill any dangerous bacteria which may still be present. In areas where the buildings are at a higher level than the reservoir which supplies them, such as those on top of hills, a communal storage unit in the form of a water tower may have to be provided.

Wells and boreholes must comply with Building Regulations although the water may be suitable for drinking without any treatment if the inlet for the water is at least 10 m below ground level. Water from roofs, springs and paved areas is all right for general purposes but would have to be filtered and purified to make it acceptable for drinking.

All pipes carrying the water supply from the company's main to the building (see figures 1 and 2) run below ground. When once on the site of the structure to be served, they become the responsibility of the building owner and may need protection depending on the type of soil in which they run and the kind of material specified. A means of isolating the supply from the building is essential in case the underground service pipe should need attention or if the building is to be left empty for some time.

The choice of pipe to be used below ground for main's water may depend upon the initial cost of the materials, pipe diameters, ease of handling, anticipated pressures and the life expectancy together with the degree of protection needed. It may also be affected by the location of the building, for example, where long runs have to be provided then the number of running connections needs to be kept to a minimum, in which case annealed copper or polythene might be the better propositions as they are available in coils of 150 m or more as against a 12 m maximum length for some of the rigid pipes.

Water Byelaws prohibit the use of wrought iron and steel, excluding stainless steel, for the supply of drinking water to buildings in areas listed in their Second Schedule, but these materials may be permitted for other uses such as fire fighting installations, demolition and construction work, agricultural buildings and irrigation systems. They also prohibit the use of lead for water services except possibly for repairs to existing pipes. Minute quantities of this material, if present in the water, can cause lead poisoning, and because water travels long distances and spreads quickly over large areas any contamination can soon become a serious health hazard. In areas where the water has been known to be plumbo-solvent the use of lead has been prohibited for a long time.

The jointing of pipes below ground must be of the screwed or manipulative compression type to ensure that the lengths cannot pull apart as may happen due to soil movement or the expansion of water on freezing when no provisions against these possibilities have been made. Water Byelaws require all metal fittings to be resistant or immune from dezincification. Solvent welded joints used with uPVC would be permitted.

Underground pipes

Anticipative problems	Considerations
1 Freezing of water in the pipes	Ensure that the pipe is placed at a depth of not less than 750 mm below ground level unless actually below a building, or suitably protected. See figure 3. *Note* Care is essential when landscaping at a later date, after the pipes have been installed, to ensure that this depth has not been reduced otherwise the protective cover is inadequate. If the depth down to the stopvalve has been increased due to a raising of the ground level then special tools will be required to reach this valve and turn it off when needed.
2 Shearing of the pipes due to **(a) movement of adhesive soils**	Ensure that the pipes are surrounded with not less than 50 mm of sand or non-cohesive fine soil in order that the existing ground can move independently of the pipe as in figure 4(a), or run the pipes through a duct of second quality of similar pipe. See figure 4(b). *Note* The volume of adhesive soils such as clay will alter depending on its water content, increasing as the soil becomes wet and shrinking as it dries out. At the same time, because of the adhesive nature of the clay it will stick to the pipes buried in it, and should these be of a soft material, such as annealed copper, the combined effect of the adhesion and the soil movement can set up stresses which eventually may cause the pipe to shear.
(b) settlement of the ground	Provide a small excess length of pipe by laying it snake fashion in the trench rather than in a straight line between two points when installing flexible pipes in made up ground. See figure 5. *Note* When ground settles it dips and if a pipe is rigidly stretched across this hollow it is subject to pressures from above only. If the pipes are of soft metal or polythene in made up ground or trenches that have been excavated too deep and then backfilled but not properly consolidated, these pressures may cause the pipes to fail. The initial excess in length will usually provide enough slack to allow for any reasonable settlement. Polythene is less likely to shear than soft metal, but failure can occur at the joints which may pull apart if non-manipulative compression fittings are used and the pipe is stretched too much.
3 Perforation or kinking of the pipe walls due to **(a) stones and flints**	Provide pipe ducts as mentioned under shearing. When the soil contains large stones and flints the pressures from above may cause any of these which may be in contact with the soft metal pipes to deform or perforate them if no protection is provided.
(b) pressure from above and around	Ensure that all flexible pipes are run through pipe ducts of a stronger material, especially where any traffic may run above and when the depth has to be less than 750 or in excess of 1300 mm.

Underground pipes *continued*

Anticipative problems	Considerations
4 **Corrosion of the pipes**	Ensure that all steel pipes, if permitted, are galvanised internally and externally. A weakness arises however when the pipes have to be cut and re-threaded to fit the particular location. Rusting will quickly occur if the cut ends are not treated before fixing. Ensure that steel and concrete pipes have an adequate protective cover in the form of a bituminous coating, wrapping with a water repellant tape, or shrouding in a plastic sleeve if used in areas where they may be attacked by chemicals in the surrounding soils.
5 **Contamination of the pipes**	Water Byelaws prohibit service pipes from passing through contaminated areas such as manholes, sewers, etc, they also state that the materials must not be susceptible to permeation by gas or other pollutants.
6 **Isolation of the supply**	Provide a high pressure stopvalve on the incoming service pipe near the site boundary. This is usually located in a brick box with a cast iron or PVC cover for access. The base of the box must be left in its natural state in order that rain water which will drain into it can seep away. See figures 2 and 6.
7 **Air locks**	Provide automatic air valves at the top of any rise in the run otherwise an airlock may occur and the flow will cease. See figure 7.
8 **Removal of any debris from the pipes**	Provide washout or drain valves at the lowest points in the installation or runs in order that any sediment can be washed out if so required. See figure 7. *Note* Water pipes below ground do not go in straight lines between two points but rise and fall with the natural contours of the ground, keeping generally between 750 and 1300 mm below the surface.

Cold water pipes above ground

These are classed as service pipes which carry the main's water either direct to appliances or to the internal cold water storage cistern, and distribution pipes which carry stored water either from the cistern or from the hot water storage cylinder to where it is required.

Generally water pipes installed above ground will differ from the underground installation in that if exposed they may have to withstand physical impact or may require special finishes in order to blend in with the general decor. Invariably they will be a collection of short runs because of the number of tapping off points and the changes in the direction necessitated by walls, floors and the positions of the units.

Capillary and compression joints considerably affect the costs and finished appearance of the work as the former are cheaper and neat but do not offer any facilities for dismantling should it be required. Other factors are the number of valves, access points, etc, needed for a satisfactory working installation and reasonably economic maintenance.

11

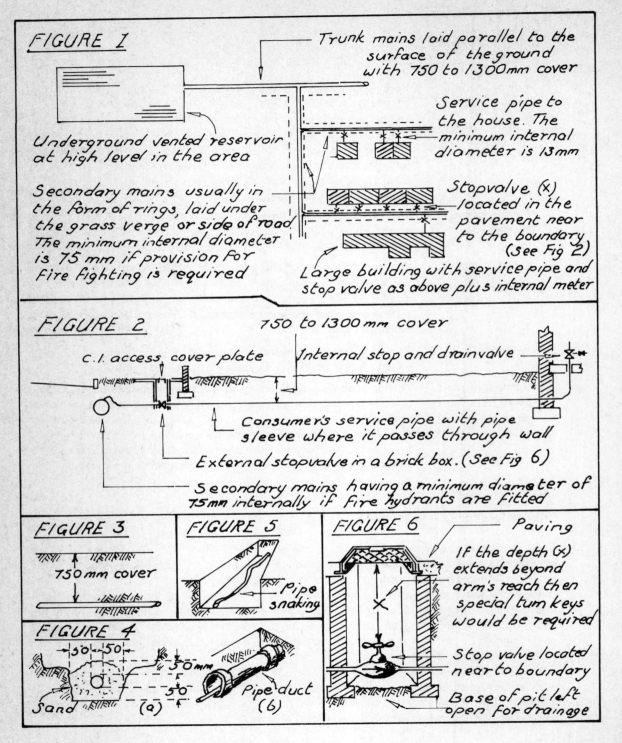

FIGURE 1

Trunk mains laid parallel to the surface of the ground with 750 to 1300mm cover

Underground vented reservoir at high level in the area

Service pipe to the house. The minimum internal diameter is 13mm

Secondary mains usually in the form of rings, laid under the grass verge or side of road. The minimum internal diameter is 75 mm if provision for fire fighting is required

Stopvalve (x) located in the pavement near to the boundary (See Fig 2)

Large building with service pipe and stop valve as above plus internal meter

FIGURE 2

750 to 1300 mm cover

c.l. access cover plate

Internal stop and drainvalve

Consumer's service pipe with pipe sleeve where it passes through wall

External stopvalve in a brick box. (See Fig 6)

Secondary mains having a minimum diameter of 75mm internally if fire hydrants are fitted

FIGURE 3

750 mm cover

FIGURE 5

Pipe snaking

FIGURE 6

Paving

If the depth (x) extends beyond arm's reach then special turn keys would be required

Stop valve located near to boundary

Base of pit left open for drainage

FIGURE 4

50 50

50 mm

50

Sand

(a)

Pipe duct

(b)

Rising main

Anticipative problems	Considerations
1 **Freezing and subsequently burst pipes**	Install the pipe in an area which is not usually subject to freezing in normal winters. These are on internal walls rather than the internal face of external walls, especially if the rooms have an east or north aspect, and near to where the water is needed so that long horizontal runs are unnecessary and kept to a minimum. Where the rising main is taken up through a garage then it would need to be lagged as this area is not heated.
	Ensure that all cold water service pipes which are carried along and up through ventilated floor spaces are lagged to protect them from the cool draughts resulting from ventilation. See figure 8.
	Lag and/or box in all pipes which have to run through areas were freezing is highly likely in normal winters, these being garages, roof spaces, external store areas and internal spaces with an east or north facing external wall.
2 **Corrosion of pipes**	Provide sleeves or wrap pipes which rise through concrete or oak floors. This permits independent movement of the pipes and the adjacent building materials, and protects metal tubes against any corrosion resulting from contact with alkali in the wet concrete or tannic acid in the oak.
3 **Internal isolation of the supply**	Provide a stopvalve to enable the supply to be turned off in cases of emergency such as when a pipe bursts, and when maintenance jobs have to be carried out, it is usually located about 150 mm off the lowest floor and must be accessible and easy to operate. See figure 8. When the valve is to be screened from view, for example in a duct or boxed in below an appliance, an access panel in the form of a hinged or lift-out plate or flush panel held in position with screws or magnetic catches must be provided. See figure 10. Where flats are connected to a common rising main then each flat must have its own high pressure stop valve located as before but just after the service pipe has been tapped off that rising main.
4 **Drainage of the pipes**	Provide a drainvalve just above the stopvalve on the rising main, or the two may be a combined. It needs to be a minimum of 150 mm off the floor in order that a hose can be attached and a small receptacle placed below the outlet to catch the drips which occur after the hose has been removed. See figure 8.
5 **Condensation on service pipes**	Avoid horizontal runs of exposed service pipe near ceilings in rooms which are heated. Where such runs are unavoidable then lagging and boxing in of the pipes is essential.
	Note Warm air in a room will rise and when it contacts the cold service pipe condensation will occur on the tube. Each time cold water is drawn off the pipe receives a fresh charge of very cold water which cools the pipe further and makes the problem worse and the dripping could be continuous.

13

Water storage

In some areas all domestic water is permitted to be taken direct from the mains to the taps, but in others, where the peak demands exceed supply at certain times of the day, or where the pressure is known to be low at peak periods of usage then water storage is essential. The minimum volume to be stored to ensure a 24 hour uninterrupted supply to all fittings can be determined by consultation with the local Water Authorities as it can vary with each area.

The pressure of the water at the taps which controls the rate of flow will depend mainly on three things, these being,

(a) the amount of head which is available. This is the actual distance measured vertically from the underside of the supply pipe outlet at the cistern down to the tap being used. This will be different for various fittings as it depends upon the height of the tap from the floor level of the room in which it is located. No head should be less than 1 m.

(b) the actual length of pipe run from the cistern to the tap.

(c) the friction or resistance to the flow set up by the pipe walls, bends, tees, etc, — which must be overcome by the force of gravity or the power of a pump. This frictional resistance varies with the length of the proposed run and the type of material used. A rule of thumb often adopted for preliminary pipe sizing was to measure the actual run of the pipe and add on one third which assumed very approximately that a 9 m run with fittings was the same as a 12 m length without. See figure 11. If the head cannot be raised or the runs reduced and the resistance is too high then the diameters of the pipes and fittings must be increased in order to obtain satisfactory flow rates. This will increase the installation costs and is a sound reason for keeping appliances near the supply pipes.

Cold water storage cisterns have two capacities, these being the nominal capacity which is the volume of water the cistern will hold before any holes for pipes are made in it, and the actual capacity which is the amount which could be drawn off if not automatically refilled. The actual capacity is about 80% of the nominal one due to the positioning of the pipes in compliance with Byelaws. See figure 12.

A ballvalve controls the flow of water into the cistern and automatically shuts off the supply when the correct water level in the unit is reached. Should this ever cease to prevent water entering when full, an overflow called a warning pipe will operate. This is a length of pipe, often short, fitted near the top of the cistern and primarily to give notice of a malfunctioning ballvalve. The Water Authorities require it to discharge in a conspicuous position and cause a nuisance so that the fault will be noticed and corrected quickly, and the resultant waste of water kept to a minimum. See figures 32 to 37.

All storage and feed cisterns, and flushing cisterns if not of the automatic type, must be fitted with ballvalves and warning pipes.

Cold water storage cisterns

Anticipated problems	Considerations
1 The capacity and need for a cistern	Check with the local Water Authorities whether or not one is required the minimum actual capacity if feeding a hot water system is 115 litres or 230 if providing water for other purposes as well. *Note* Usually buildings in flat country, those on high ground and where customer's requirements are above average or involve industrial processes are the most likely to need them.

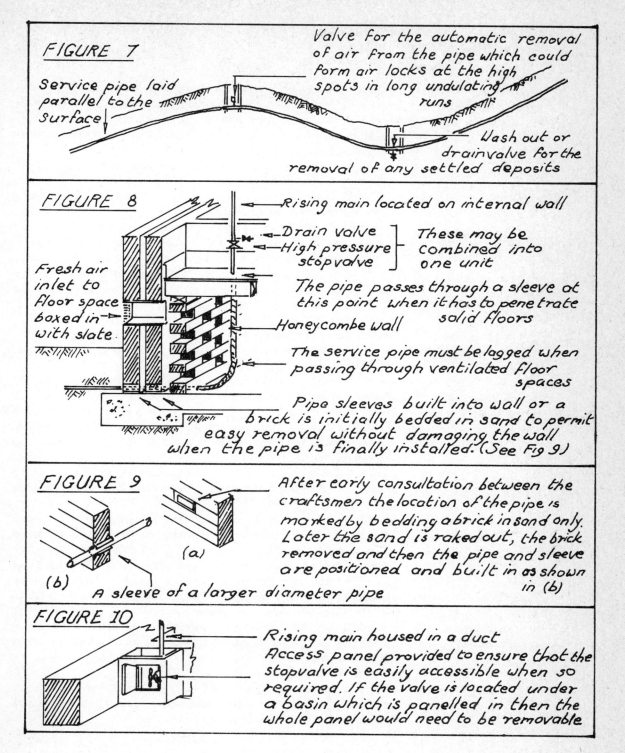

FIGURE 7

Service pipe laid parallel to the surface

Valve for the automatic removal of air from the pipe which could form air locks at the high spots in long undulating runs

Wash out or drainvalve for the removal of any settled deposits

FIGURE 8

Rising main located on internal wall

Drain valve
High pressure stopvalve
} These may be combined into one unit

Fresh air inlet to floor space boxed in with slate

The pipe passes through a sleeve at this point when it has to penetrate solid floors

Honeycombe wall

The service pipe must be lagged when passing through ventilated floor spaces

Pipe sleeves built into wall or a brick is initially bedded in sand to permit easy removal without damaging the wall when the pipe is finally installed. (See Fig 9)

FIGURE 9

(a)

(b)

A sleeve of a larger diameter pipe

After early consultation between the craftsmen the location of the pipe is marked by bedding a brick in sand only. Later the sand is raked out, the brick removed and then the pipe and sleeve are positioned and built in as shown in (b)

FIGURE 10

Rising main housed in a duct
Access panel provided to ensure that the stopvalve is easily accessible when so required. If the valve is located under a basin which is panelled in then the whole panel would need to be removable

Cold water storage cisterns *continued*

Anticipated problems	Considerations

2 Access to and the replacement of cisterns

Provide an access hole with measurements between the ceiling hatch suports at least 150 mm more than the nominal length and width of the storage cistern. Allowances must be made also for loft ladders if proposed as these considerably restrict the space available when down to provide access.

Note As cisterns do not last as long as the building then replacements will be needed periodically.

Check joists as parts may have to be removed and trimmers of a larger section fitted. See figure 13.

Ensure that the hatch is near a support wall as this will provide a rest for the ladder until it is open, when the ladder can be passed through.

When cisterns are sited in cupboards ensure that the space between the doorstops is wide enough to permit a replacement at a later date.

Note Often these days cisterns are put into spaces before the structure is finished with little or no thought being given to the problem of replacements.

3 Weight of the cistern when full

Locate the cistern over a wall rather than mid-span over the joists or spread the load by the use of bearers. The minimum weight of a cistern feeding a hot water cylinder is approximately 230 kg.

Note 1000 litres = 1000 kg = 1 tonne = 13 m (approx).

4 Pressure and available head

Ensure that the minimum head is never less than 1 m.

Note Cisterns located in roof spaces usually present no problems but difficulties may arise for the supply to fixed high level shower heads on the top floors. The pressure can be raised by placing the cistern on a stollage. See figures 11 and 14.

When buildings have flat roofs or cisterns have to be sited in cupboards, etc, then the water supply to the shower heads may have to be pump assisted. See figure 15.

5 Malfunctioning ballvalves

Provide a minimum of 300 mm clear space above the top of the cistern in order that maintenance to the ballvalve may be carried out. Washers wear and the float and valve need replacing periodically. See figure 16.

When the hot water system vent and expansion pipe discharges over the cold water cistern then the 300 mm may have to be extended, dependant upon the available head to the hot water installation. See *Hot Water systems with central boiler,* problem 2, page 50.

6 Space around the cistern

Provide a minimum of 300 mm on all sides of the cistern for the fixing of the inlet and outlet pipes and the insertion of tank lagging. In large tanks this may have to be increased considerably if connections are difficult to reach. See figure 16.

7 Warning pipes

Provide a warning pipe for all cold water storage cisterns controlled by ballvalves and ensure that it discharges at a point which will cause a nuisance. See figures 16 and 32 to 37 inclusive.

16

FIGURE 11

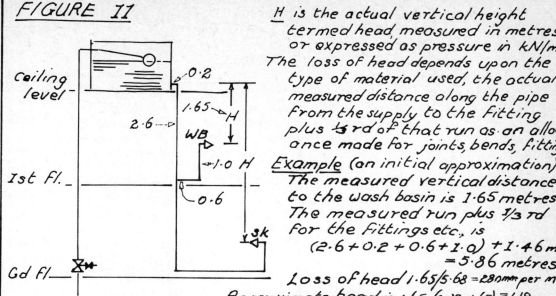

ceiling level

1st fl.

Gd fl.

0.2
2.6
1.65 H
WB
1.0 H
0.6
Sk

H is the actual vertical height termed head, measured in metres or expressed as pressure in kN/m^2.
The loss of head depends upon the type of material used, the actual measured distance along the pipe from the supply to the fitting plus ⅓rd of that run as an allowance made for joints, bends, fittings

Example (an initial approximation)
The measured vertical distance to the wash basin is 1.65 metres
The measured run plus ⅓rd for the fittings etc., is
$(2.6 + 0.2 + 0.6 + 1.0) + 1.46m$
$= 5.86$ metres
Loss of head $1.65/5.68 = 280mm$ per m
Approximate head is $1.65 - (0.28 \times 1.65) = 1.19m$

Note:- The value of H for the sink will be greater because of the bigger drop. Minimum value is 1 metre

FIGURE 12

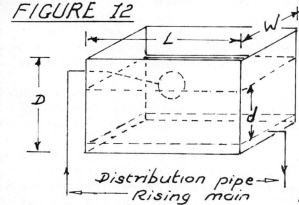

W
L
D
d

Distribution pipe →
← Rising main

The nominal capacity is calculated by using the actual dimensions of the cistern which is
$$L \times W \times D = X m^3$$
Note:- $1 m^3 = 1000$ litres
The actual capacity is the volume of water which may be drawn off before refill
i.e. $L \times W \times d = X m^3$
The actual capacity could be 20% less than the nominal one

FIGURE 13

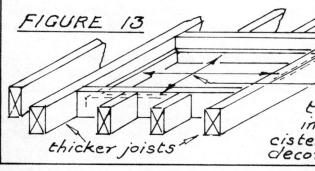

thicker joists

The clear space measured between the supports for the hatch cover must be greater than the nominal capacity dimensions otherwise there will insufficient room to pass the cistern through and the finished decorations could be damaged

Cold water storage cisterns *continued*

Anticipated problems	Considerations
8 Freezing of the stored water	Provide lagging to the top and all sides of the cistern but it must not be airtight. Lagging beneath may be omitted if the ceiling insulation has already been installed and it is not on a stollage. See figure 16. Provide a bend and extension to the warning pipe within the cistern so that it terminates below the surface of the water, and also a hinged flap on the external end which will allow water to flow out but prevent cold draughts blowing in. See figure 17.
9 Provision of drinking water	Provide one tap direct off the service pipe in situations where all the other cold water taps are fed from the storage cistern. See figure 18. Some of the small industrial and commercial buildings may need to have drinking fountains provided.
10 Leakage from cold water cisterns	Provide a metal safe below the cistern to catch any drips resulting from leaks, and connect this to the warning pipe ensuring that the tee connection is below the level of the safe. This helps to eliminate the inconvenience and possible expensive repair work caused by the leak from above which could occur when the building is unoccupied. On larger buildings an asphalt safe on concrete floor extending over the whole tank-room may be necessary.
11 Use of more than one cistern	Where two or more cisterns are used in place of one because of possible height restrictions or space limitations then the combined capacity of them must not be less than if one had been used. Any draw-offs must be from all cisterns simultaneously otherwise the water will remain in those not directly connected to the delivery pipe and eventually turn green.
12 Choice of materials	Galvanised iron and steel cisterns are strong and rigid, but susceptible to rust, and electrolysis which can lead to rusting. Electrolitic action is less likely to occur in cold water cisterns than with hot water tanks in circulatory heating systems. Copper cisterns are small, but where larger ones are required then the copper sheet is supported on timber stiffeners to give the sheet rigidity. Copper must not be used with galvanised steel pipes because of the possibility of electrolysis destroying the zinc in the galvanising. Asbestos cement cisterns are heavy and tend to be brittle but are not subject to electrolysis. Care is needed when perforating the material for pipes as it may spall, making it more difficult to form watertight connections. Polythene, polypropylene and glass fibre are all very light in weight so easy to handle and are not affected by electrolysis. Polythene cisterns may be folded which is an advantage if the access is too small but of no consequence if adequate room had been left initially.
13 Prohibited materials	Lead cisterns are not permitted, and in certain areas stated in the Water Byelaws Second Schedule cisterns of iron and steel, excluding stainless steel, may not be used.

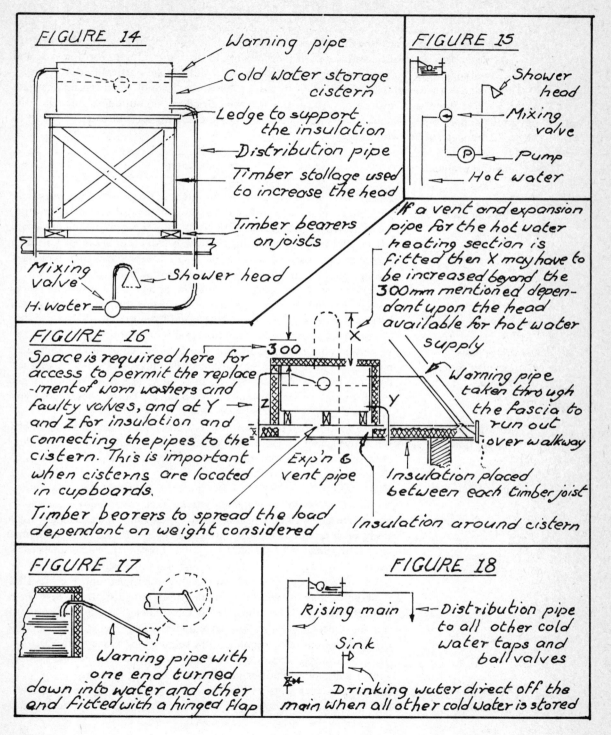

FIGURE 14

Warning pipe

Cold water storage cistern

Ledge to support the insulation

Distribution pipe

Timber stollage used to increase the head

Timber bearers on joists

Mixing valve

H. water

Shower head

FIGURE 15

Shower head

Mixing valve

Pump

Hot water

If a vent and expansion pipe for the hot water heating section is fitted then X may have to be increased beyond the 300mm mentioned dependant upon the head available for hot water supply

FIGURE 16

Space is required here for access to permit the replacement of worn washers and faulty valves, and at Y and Z for insulation and connecting the pipes to the cistern. This is important when cisterns are located in cupboards.

Timber bearers to spread the load dependant on weight considered

300

X

Z

Y

Exp'n & vent pipe

Warning pipe taken through the fascia to run out over walkway

Insulation placed between each timber joist

Insulation around cistern

FIGURE 17

Warning pipe with one end turned down into water and other end fitted with a hinged flap

FIGURE 18

Rising main

Sink

Distribution pipe to all other cold water taps and ball valves

Drinking water direct off the main when all other cold water is stored

Cold water distribution pipes

These vary from the service pipe in that instead of taking the main's water at high pressure up to the storage cistern they carry the stored water from it and distribute it in different directions to the cold water taps and ballvalves located in the building. In distribution pipes for domestic and low rise buildings the water pressure is low when compared with the main's generally.

The need to run across and around a building raises problems due to door openings and obstacles such as walls, floors and joists, etc, vertical runs may be carried up in continuous ducts but the horizontal supply pipes are not so easy to conceal. If they are left exposed then it is difficult to blend their appearance in with the general finishes.

Anticipative problems	Considerations
1 **Passing pipes through walls**	Ensure that there is early consultation between the brickwork and plumbing craftsmen regarding the approximate position where the pipes are expected to pass through the walls. If at these points the bricks are left in sand only, then all the plumber has to do to provide a hole is to rake out the sand and remove the brick. The pipe can then be passed through and the brickwork made good. See figure 9. *Note* Whilst today many holes may be drilled if the plant and power is available, and the pipe diameter not too large, the provision of bricks in sand prevents any need for cutting into the brickwork and possibly causing damage and undue vibration to the structure.
2 **Pipes passing through concrete floors**	When pipes are too large for holes to be provided by drilling then they must be preformed in the wet mix. Drilling small holes in finished concrete may be difficult because of the presence of the reinforcement which cannot be seen in the finished product. This applies to both pre-cast and cast 'in situ' types of concrete.
3 **Housing pipes in timber floors**	Ensure that all pipes which run below the level of the timber floor can run between and parallel to the joists. This may mean a reduction in the span of some joists and the use of a trimmer to support the ends of them. See figures 19(a) and (b). *Note* Any notching of the joists must not be tolerated as this reduces the designed effective depth and strength of them. Also when floorboards are being fixed it is not possible to see the pipe and a nail might easily perforate one, and this fault may not be obvious until some time later. The leak can be difficult to find and repair, by which time costly damage may have occurred to some of the finishes below. *Note* Because of the short distances between joists and the fact that rigid pipes may be up to 6 m long it is not possible to pass them through holes drilled in the sides of the timber. The alternatives are to run the pipes above the floor avoiding crossing openings, or running below the ceiling and dropping to the required level. See figure 20. *Note* The suggestion of a false ceiling over the whole room to conceal a few small pipes cannot be justified on economic grounds.
4 **Doors and openings generally**	Ensure that pipe runs within a room terminate, if above floor level, before reaching any openings otherwise they will either run up and

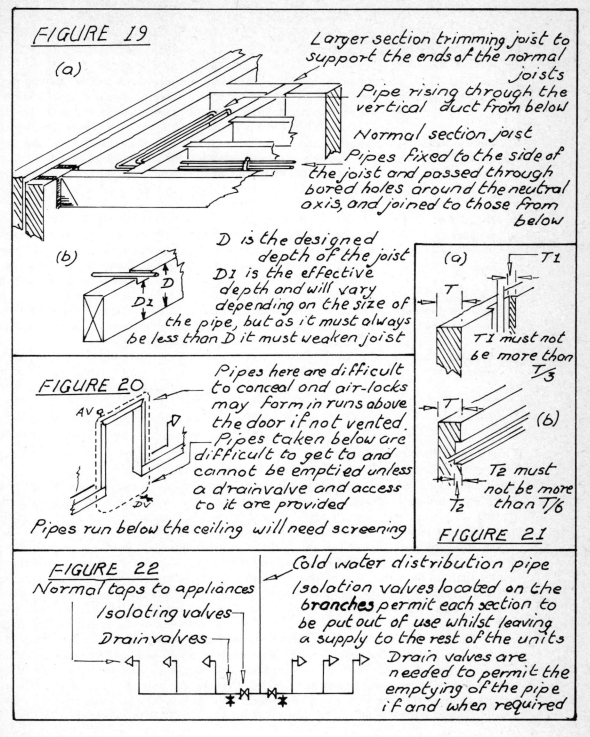

FIGURE 19

(a)

Larger section trimming joist to support the ends of the normal joists

Pipe rising through the vertical duct from below

Normal section joist

Pipes fixed to the side of the joist and passed through bored holes around the neutral axis, and joined to those from below

(b)

D is the designed depth of the joist
D1 is the effective depth and will vary depending on the size of the pipe, but as it must always be less than D it must weaken joist

FIGURE 20

AV

DV

Pipes here are difficult to conceal and air-locks may form in runs above the door if not vented.
Pipes taken below are difficult to get to and cannot be emptied unless a drainvalve and access to it are provided

Pipes run below the ceiling will need screening

(a)

T1

T

T1 must not be more than T/3

T

(b)

T2 must not be more than T/6

T2

FIGURE 21

FIGURE 22

Cold water distribution pipe

Normal taps to appliances

Isolating valves

Drainvalves

Isolation valves located on the **branches** permit each section to be put out of use whilst leaving a supply to the rest of the units

Drain valves are needed to permit the emptying of the pipe if and when required

Cold water distribution pipes *continued*

Anticipative problems	Considerations
	around it, providing reasons for periodic loss of supply resulting from possible air locks unless airvalves and vents are provided, or run below the opening necessitating drain valves and access panels at the bottom of the dip in order that it can be emptied without mechanical assistance.
5 **Pipes in wall chases**	Ensure that any chase cut to house a pipe but not filled in again does not adversely affect the stability of that wall. Buildings Regulations require that the maximum depth of a vertical chase must be not more than one third of the wall thickness, and horizontal ones not more than one sixth. See figures 21a and b.
6 **Isolation of supplies**	Provide low pressure gatevalves to cut off the supply when maintenance such as changing washers or the replacement to taps or ballvalves is needed. There must be one on the main distribution pipe and if there are a number of branches with several appliances on each, then one is required on each branch in order that it may be isolated and leave the service to the rest of the building still in operation. See figure 22.
7 **Draining down of the branches**	Provide drainvalves on all branches fitted with isolating valves, locating them at the lowest part of each branch, between the gatevalve and the actual appliance being served. See figure 22.
8 **Freezing of the pipes**	Lag all pipes running in ventilated floor spaces and those in roof spaces above the thermal insulation placed between the joists. Avoid running pipes near the eaves as not only is freezing more likely here but the pipes are not so accessible.

Flow controls for cold water

These controls are classed as (i) taps, that is fittings connected to the end of a pipe run and from which water is directly obtained by various mechanisms, (ii) valves which are located within the run and which, when once adjusted, are not altered except when emergencies occur or maintenance has to be undertaken, and (iii) automatic controls such as ballvalves.

Taps generally are in one of two groups these being bib taps which have a horizontal inlet as shown in figure 23, and are fitted direct to a wall fitting, and pillar taps which have a vertical inlet and are fitted to the appliance as in figure 24. Also they may be one of two types, those which are turned by hand but actually open due to water pressure alone, and are fitted to service pipes so subject to high pressures, and those which are provided with some physical assistance to lift the washer off its seating as in figure 25(c) and are used on low pressure supply pipes connected to the cistern. Should a high pressure tap be fitted on a low pressure pipe, especially where the head is less than 2 m, and the occupants turn off the water and close the building for a few weeks whilst on holiday, the tap washer could become stuck on its seating. Even after the supply is reinstated considerable delay and inconvenience could result before any water is obtained from the tap because although the pressure is enough for normal daily use it is quite inadequate to lift a washer which has become stuck resulting from being clamped down for those few weeks.

Taps can be operated in various ways. The most common method is by pressing a knob or turning a handle. For the disabled, and for special use in clinics and hospitals, other methods such as levers which can be rotated at 90° by hand, wrist or elbow, exist.

The usual sizes for taps are 15 mm for wash basins, 22 mm for baths and 15, 22 and, in some cases, 28 mm for sinks.

Taps

Anticipative problems	Considerations
1 Materials and finishes	**Brass** This is an alloy of copper and zinc and when carrying certain types of water is subject to cavitation or dezincification. As the zinc is removed small channels form in the seating of the tap which permit water to flow even when the washer is clamped down tight thus causing continual dripping. See figure 25(a). The finishes on brass may be natural, polished or chromium plated. **Gun metal** This is a bronze or alloy of copper and tin and although more expensive than brass is not subject to cavitation. The finishes of this material is the same as for brass. **Acrylics** The whole or part of the tap may be in plastics and these have an advantage over metal in that the finishes may be gold, chromium or in a matching colour to the appliance. Acrylic taps are not as robust as metal ones, but are not subject to cavitation.
2 Water waste and economic use	In public establishments, factories and schools the taps are liable to be left running after use. This constitutes considerable waste of water and contravenes Water Byelaws. It also results in extra high water bills where the supply is on a meter. Spring operated, nonconcussive, automatic closure type taps are advisable in these circumstances. A spring below the operating knob will shut off the supply once the hand pressure on the knob has been removed. It is economical in the use of water but not suitable for domestic needs. Spray taps also conserve water but do not have the advantage of the automatic shut off. Other taps reputed to be economic in water usage are those which appear as a pillar tap but supply hot and cold water from the same unit. As the hot and cold pressures must be reasonably identical they must have the same head. As the hot and cold are combined in one unit it leaves a position on the appliance for another fitting which might be a soap dispenser fed from a central tank above.
3 Maintenance to taps	Provide isolating valves on the main distribution pipe and, if non-domestic, all branches. These valves are closed should a tap need to be changed or a washer on a tap replaced. Where installations do not have these valves, or the closing of them is not desired then 'Supataps' may be advisable. As the jumper and washer are removed from the outlet of the tap another small stop within the unit slides down and seals off the unit even if the supply is not turned off. This avoids the necessity of isolating other appliances when attending to a tap but isolation would be essential should the whole tap have to be replaced.

23

Taps *continued*

Anticipative problems	Considerations

4 External water supplies

Where water is required for the washing of cars and watering of gardens, etc, then provide taps with threaded outlets to take a hose union. See figure 23. It would be advisable to house these units in a weatherproof box which, together with some insulation will provide the necessary frost protection during winter.

Valves

These are fitted in the flow path rather than at the end of a run as are taps, and they fall generally into four groups. See figure 31.

1 High pressure stopvalves which have loose jumpers and washers, as in figure 25(b) and are fitted on service pipes to isolate the mains from the appliances or from the establishment as a whole.

2 Low pressure gatevalves which are fitted on distribution pipes running from the cold water storage cistern, and used to isolate the stored water supply during maintenance work on distribution runs.

3 Drain valves which are located at the bottom of any drops where the pipes run below the level of the lowest tap or ballvalve it feeds. Generally any low level isolating valve is accompanied with a drain valve and with some repairs, for example a leaking capillary joint, the emptying of the pipe is an essential preliminary operation.

4 Ballvalves. These are operated automatically by a float which closes the valve as the water rises in the cistern and opens it again as the level drops. They are fitted in all cold water storage or feed cisterns and on all hand operated flushing cisterns.

Anticipative problems	Considerations

1 Isolation of the supply internally and externally

Provide high pressure stopvalves on all service pipes as shown in figure 31. The locations are

(a) near the site boundary as in figure 6.

(b) near the entry point of the mains into the interior of the building as in figures 2 and 8.

(c) just prior to each storage and feed cistern when more than one are connected direct to the common rising main.

(d) on branches supplying mains water direct to the appliances where the isolation of supplies to the whole building for repairs to one tap cannot be tolerated or is not recommended.

Note Ensure that the arrows cut or moulded on the valve body, as shown in figure 26, point in the direction of the proposed flow otherwise instead of lifting the washer off its seating the flow of water will hold it down and prevent any flow even though the valve appears to be open.

FIGURE 23

Bib tap

Flow

These taps may have

(a) crutch, cross, wheel, star, lever metal or moulded acrylic operating control

(b) a natural, polished, chromium plated metal or coloured acrylic body which may be as shown or inclined

(c) plain outlets with anti-splash insets or threaded for a removable hose union

FIGURE 24

Pillar tap

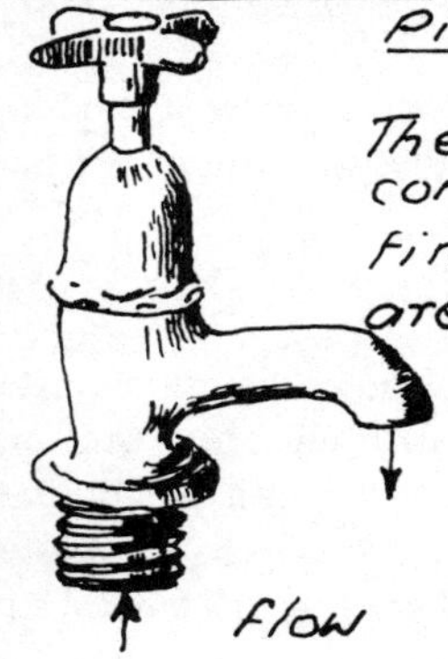

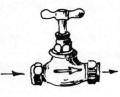

Flow

The operating control, materials, finishes and outlet are as in Figure 23 but the shapes and patterns vary considerably more than above including mixing valves

FIGURE 25

Groove formed in the seating (a) resulting from the type of water used thus allowing water to flow even when the tap is turned off.

(b)

Loose jumper and washer in a high pressure tap

Loose jumper with a grub screw engaged in groove which lifts the jumper off it's seating as the hot tap is turned on (c)

FIGURE 26

High pressure stopvalve

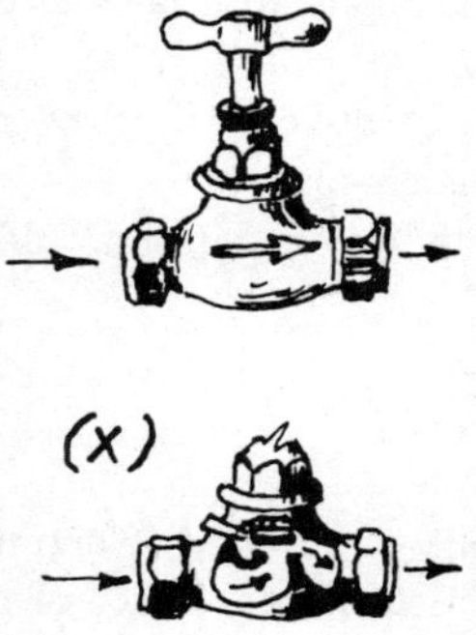

(X)

Note:- The arrow cut or cast on the body of the valve indicates the direction of the flow. If the valve is fitted the wrong way round then the water will flow above washer (X) and hold it down on to it's seating making it a non-return valve and stopping flow

FIGURE 27

Low pressure gate valve

This valve has no washers to be replaced. As the control is turned so is a worm spindle which causes a metal gate to slide across the orifice, opening or closing the valve as required

worm thread

metal tapered gate.

Valves *continued*

Anticipative problems	Considerations
2 Isolation of the stored water supply	Provide gate valves on the common cold water distribution pipe running from the cistern to the taps. When turned off this isolates the supply to permit repairs such as replacing washers on taps when needed to be carried out. See figures 27 and 31. Provide gatevalves on branches supplying two or more appliances. This enables maintenance to be carried out on that branch without having to isolate all the units in the building at the same time. See figures 22, 27 and 31.
3 Drainage of all or parts of the system	Provide drainvalves at the bottom of each drop. *Note* An open tap can only drain a pipe down to same level as itself. Any pipe below the tap must remain full unless a drainvalve is fitted. See figures 30 and 31. Provide drainvalves to all branches fitted with isolating valves and locate them at the lowest point between the gatevalve and the outlet on the appliance. See figures 22 and 31. Ensure that no drainvalve goes below ground level or is in a position likely to be flooded.

4 Automatic water supply to cisterns	Provide ballvalves to all cold water storage, feed and flushing cisterns, ensuring that it is suitable for the pressure involved. There are various patterns and the size of the inlet will differ with the anticipated pressure the valve has to close against. The higher the pressure to be resisted is the smaller the orifice in the valve will be. Types of ballvalve available are,
	    (a) Portsmouth, see figure 28 and Croydon. The former is the most popular and the quieter of the two.
	    (b) Diaphragm valve. See figure 29. All of the working parts are dry so there is less likelihood of deposits resulting from the evaporation of water causing the valve to seize up. Acrylic versions are available but are not as robust as the metal ones.
	    (c) Equilibrium ballvalve. These are more expensive than others but are best where the pressures vary considerably each day.
	Note Some valves have the pressure indicated by letters cut or moulded on the valve body.
	    HP for high pressure — 140 m head approx.
	    MP for medium pressure — 70 m head approx.
	    LP for low pressure — 30 m head approx.
	The head is the pressure against which the valve must close. When it is very low then a fullway valve, which has an inlet and an outlet of the same diameter may be used, otherwise if the orifice is too small then time engaged in refilling would be quite unacceptable.
	
	Provide a minimum space of 300 mm above the valve to permit repairs or replacements of floats, washers or even the valve itself.

FIGURE 28

Portsmouth type ballvalve

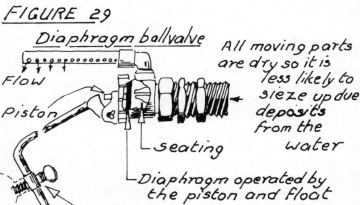

Flow

Removable cap for releasing the piston and washer. The latter needs to be replaced periodically

Float to be attached to the valve

Seating which may be of brass, bronze or nylon. The size of the orifice varies with the degree of pressure anticipated, and the smaller the hole the greater the pressure the valve resist

FIGURE 29

Diaphragm ballvalve

Flow

Piston

All moving parts are dry so it is less likely to sieze up due deposits from the water

seating

Diaphragm operated by the piston and float

Adjustable float fixing device

FIGURE 30

Drainvalve

washer

Flow

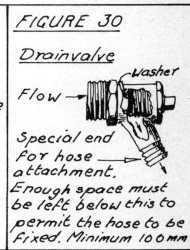

Special end for hose attachment. Enough space must be left below this to permit the hose to be fixed. Minimum 100mm

FIGURE 31

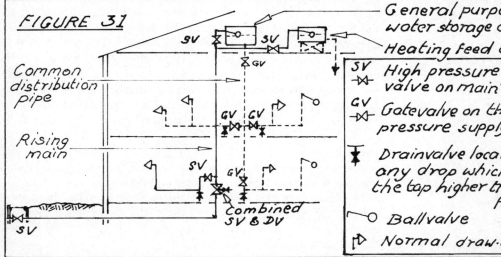

General purpose cold water storage cistern

Heating feed cistern

Common distribution pipe

Rising main

Combined SV & DV

SV High pressure stop-valve on main's supply

GV Gatevalve on the low pressure supply pipe

Drainvalve located on any drop which has the tap higher than it's feedpipe

Ballvalve

Normal draw-off tap

Valves *continued*

Anticipative problems	Considerations
5 Flooding due to the malfunctioning	Provide warning pipes as before described, minimum diameter 18 mm, to all units fitted with ballvalves. The discharge point for this pipe will depend upon the location of the room and the position and height of the cistern in that space. It must be conspicuous so that it can be readily seen and possibly be a nuisance when discharging so that the fault will be put right quickly and any subsequent wastage of water kept to a minimum. See (a), (b) and (c) following and figures 32 to 37.
(a) Flushing units fitted near to or on external walls	When cisterns are near to or on external walls then the warning pipe may pass through that wall and project for about 50 mm. It does not matter if the cistern is at high or low level but if the latter then the projecting pipe may be a hazard to persons walking by. See figure 32.
(b) High level flushing cistern in a room without external walls	The warning pipe may discharge over the appliance but must terminate at a height of not less than 150 mm above the rim of it as in figure 34. It could be extended to reach an external wall but it must not rise above the level of the outlet from the flushing cistern. See figure 35. It may be made to discharge over the pan provided the clearance is not less than 150 mm so to position the terminal point low down would not be practical as the seat could not be raised. It might be better fitted as shown in figure 33.
(c) Low level flushing cisterns in areas without external walls	Generally these are too low to be able to discharge over an appliance, especially as the pipe must not rise and provide the necessary clearance in order to comply with the Water Byelaws. Special fittings are available for high or low level units in domestic premises which permit the warning pipe to combine with the bath overflow. See figure 36.
	Where the above facility is not applicable or the other methods mentioned possible then the warning pipe must discharge over the floor which must of course be of concrete and be laid to fall to a trapped gulley to carry away the waste water. The gulley must not be positioned directly under the end of the pipe otherwise no nuisance will be caused, but as shown in figure 37. This method is not advisable if the floors are to be heated as the water seal in the drain trap will evaporate, and drain smells and gases will permeate into the room. Where underfloor heating is required then high level cisterns are preferable, possible concealed, with provisions for the warning pipe terminal only to be exposed as shown in figure 33.
5 Siphonage	The ballvalve must be fixed firmly to the side of the cistern at such a height as to clear the water level when closed. Silencer pipes extending from the inlet down into the water are prohibited. *Note* The above two conditions are required by Water Byelaws to prevent the siphonage and backflow of the stored water from the cistern to the main supply, and any spread of bacteria which could result from it becoming a health hazard.

28

FIGURE 32

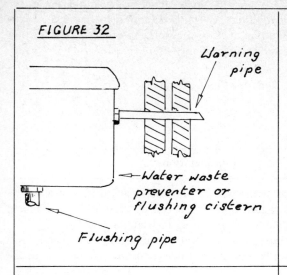

Warning pipe

←Water waste preventer or flushing cistern

Flushing pipe

FIGURE 33

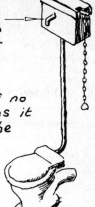

Warning pipe turned down to discharge over the appliance

Note:-
This would be of no use lower down as it would prevent the movement of the lid

FIGURE 34

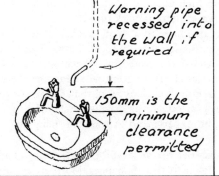

Warning pipe recessed into the wall if required

150mm is the minimum clearance permitted

FIGURE 35

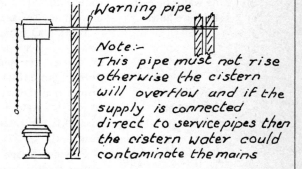

Warning pipe

Note:-
This pipe must not rise otherwise the cistern will overflow and if the supply is connected direct to service pipes then the cistern water could contaminate the mains

FIGURE 36

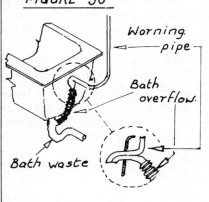

Warning pipe

Bath overflow.

Bath waste

FIGURE 37

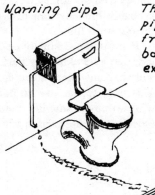

Warning pipe

The cistern and warning pipe could be screened from view and just the bottom end of the pipe exposed

The gulley could be a common one taking the discharge from several warning pipes, and located in some central position

Note Flushing valves are available but are not popular in this country probably because of the large storage cistern required and the fact that the flush, 13.5 litres in four seconds is higher than we use normally. The new Water Byelaws requiring dual flush cisterns in order to conserve our water will possibly reduce the demand for these units further.

Wells and boreholes

Anticipative problems	Considerations
1 **Continuity of supply**	Ensure that the base of the well or borehole is at least 2 m below the level of the lowest water table which can be reasonably determined otherwise it will probably dry up during each summer period. Where wells or boreholes are used for industry, for example as a water supply for bottle washing in a dairy, ensure that the pumps are duplicated.
2 **Contamination of the water in wells and bore holes**	Provide a weathered concrete or paved area not less than 1200 mm wide with a 150 mm high upstand around the well at ground level to prevent any surface water running back into it. Provide an impervious liner to a depth sufficient to prevent surface or near surface contamination affecting the water. The deeper the well the better the quality of the water in it because it has had to pass through more filter material before eventually reaching the sump.
3 **Providing the water where required**	Ensure that the actual pumps are not more than 7.5 m above the lowest anticipated water level (although the driving mechanism may be), otherwise difficulty may be experienced at times in being able to start the flow. Provide storage cisterns for at least 24 hours usage at high level to permit gravity feed down to the appliances. Mechanical pumps need to be float operated to ensure the correct water level in the cisterns. If manually operated pumps are used an overflow pipe arranged to discharge over an appliance near the operator is required as the cistern is probably hidden from his view, and any discharge will warn him that the cistern is full. If no pump is employed ensure that the well is covered when not in use, that a stand for any buckets 150 mm above ground level is provided and that any mechanism is so geared that it can be used by a female. This may appear old fashioned, but to be in a rural area and cut off with no power supplies is not unheard of even in these modern times, especially during gales in winter.

General purpose water from springs, roofs and paved areas

Spring water may be drinkable at source but may dry up during the summer if several weeks pass with no rain. Water from roofs or paved areas would need to be treated before it became of drinkable quality. If these are the only source of water supply provide a storage tank with a capacity of not less than that required for six weeks usage. A separator and/or filter will be necessary to ensure that the initial flow from roofs or paved areas which carries the bulk of any foreign matter is diverted away or filtered before storage commences.

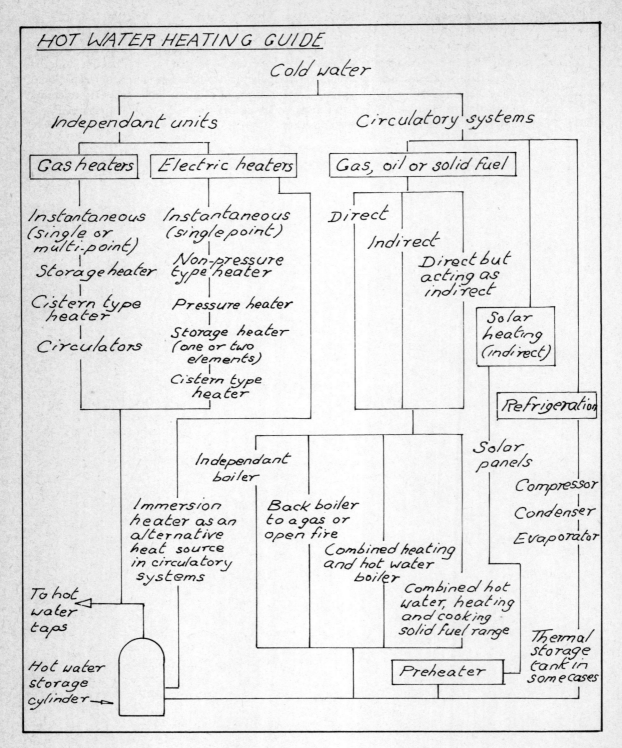

HOT WATER HEATING GUIDE

2 Hot water heating

Water for general domestic use may be heated by gas or electric factory-made independent heaters, or by incorporating a gas, oil or solid fuel boiler with a circulatory system using gravity flow or pumped hot water to transfer the heat from source to the water to be used.

The location of the heating unit will vary with the model chosen, the fuel used and the system of circulation employed. Other items to be considered are the need for a flue, possible fuel storage, automation, ventilation, economy, compliance with Building Regulations and Water Byelaws, and space around the unit to permit easy and proper servicing.

Gas fired water heaters

These units are fed direct either from the service pipe with main's water or from the feed pipe run from the cold water storage cistern. The heater may have a traditional flue or, if it is one of the room sealed types, then a balanced flue.

Anticipative problems	Considerations
1 Fuel supply	Provide for a gas service pipe and a meter. The latter will include the main isolating gas valve which must be easily accessible at all times. The meter may be located internally as would coin operated types, or in the outer leaf of the cavity wall which eliminates the need for the gas man to enter the premises to take a meter reading. See figure 38.
2 Choice of heater to be used	**1 Instantaneous water heaters** These have no storage facilities but heat the water as it flows through the unit, so gas is used only when hot water is needed. There are single and multipoint versions available according to whether it has to serve one tap or several appliances including a bath. They are good for intermittent use but unsatisfactory for situations where demand is high and constant. See figures 39 and 40.
	2 Storage heaters These may be classed as
	(a) *those with an integral heater* Normal convection currents within the unit are restricted due to small holes in the false base, causing the water to stratify and heat from the top of the unit and is usually available in a short space of time even after a previous large draw-off. The flame burns continuously but the rate of burning will vary as the gas supply will be thermostatically controlled. See figure 41. Provide also a vent and expansion pipe.
	(b) *Gas circulator* See figure 42. A small gas heater is either attached to a hot water storage cylinder or tank, or installed as a separate unit heating an independant cylinder. The water is heated as described in circulatory systems, the unit acting as a small boiler.

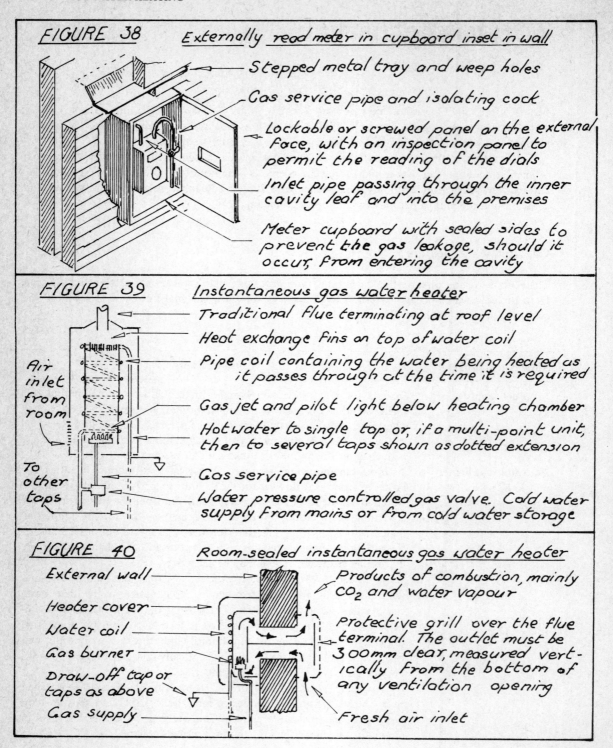

FIGURE 38   Externally read meter in cupboard inset in wall

Stepped metal tray and weep holes

Gas service pipe and isolating cock

Lockable or screwed panel on the external face, with an inspection panel to permit the reading of the dials

Inlet pipe passing through the inner cavity leaf and into the premises

Meter cupboard with sealed sides to prevent the gas leakage, should it occur, from entering the cavity

FIGURE 39   Instantaneous gas water heater

Traditional flue terminating at roof level

Heat exchange fins on top of water coil

Pipe coil containing the water being heated as it passes through at the time it is required

Gas jet and pilot light below heating chamber

Hot water to single tap or, if a multi-point unit, then to several taps shown as dotted extension

Gas service pipe

Water pressure controlled gas valve. Cold water supply from mains or from cold water storage

Air inlet from room

To other taps

FIGURE 40   Room-sealed instantaneous gas water heater

External wall

Heater cover

Water coil

Gas burner

Draw-off tap or taps as above

Gas supply

Products of combustion, mainly CO_2 and water vapour

Protective grill over the flue terminal. The outlet must be 300mm clear, measured vertically from the bottom of any ventilation opening

Fresh air inlet

34

Gas fired water heaters *continued*

Anticipative problems	Considerations

3 Water supply and pressure

(c) *Cistern type heaters* See figure 43. These have the cold water feed cistern combined with the unit thus avoiding the need to have one in the roof space. Whilst this may appear to be the solution for low buildings with flat roofs the lack of pressure due to the small head could be a disadvantage. See next item.

If an instantaneous water heater is specified then provide a feed cistern in the roof space and a feed pipe from the cistern to the heater. Alternatively, and subject to the approval of the Authorities, connect the heater direct to the mains, having first ascertained that the water supply and pressure is adequate at all times.

Note As the temperature of the water drawn off is dependant on the gas valve which is operated by water pressure, it is essential that this pressure does not vary if the temperature required is to remain constant. For this reason it is advisable for all units to obtain their water from a feed cistern which provides a constant head, especially in areas where the demand is high and the water pressure fluctuates throughout the day. See figure 44.

If storage heaters or circulators are specified then provide a cold water storage cistern at high level in the roof space. In the case of buildings with flat roofs then the cistern types of heater may be provided but it must be wall mounted above the level of the appliance in order to obtain the necessary head at the taps. See figure 45.

Note Storage cisterns may be located on roofs but must be suitably protected from the weather.

4 Location of the heating units.
See also hot water distribution pipes

Instantaeous water heaters may be installed above the level of the appliance, or below any working surfaces provided that there is adequate ventilation and provision for the flue. A model which is not room sealed should have the flue enclosed in an approved service duct.

Locate the instantaneous heater nearest to the appliance requiring hot water most frequently.

Note As the water flows so the gas will burn, and if there is a long distance between the tap most used and the heater then the pipe will be full of hot water each time the tap is turned off. This will cool and result in not only wastage of water but also of the gas just used.

Storage heaters and circulators, excluding the cistern type, may all be floor mounted.

All room-sealed units must be positioned on an external wall or back on to a duct which will provide fresh air to the burners as well as removing all products of combustion. They may be floor or wall mounted.

5 Removal of the products of combustion

Provide flues for all water heaters excluding those specifically mentioned in Building Regulations M9. These may be of circular, square or rectangular section provided that in the last instance the ratio of the

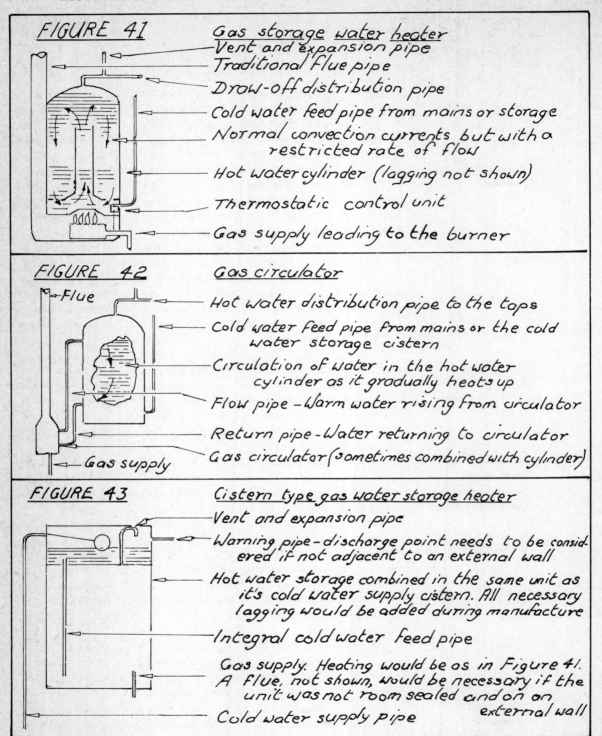

FIGURE 41   Gas storage water heater

Vent and expansion pipe

Traditional flue pipe

Draw-off distribution pipe

Cold water feed pipe from mains or storage

Normal convection currents but with a restricted rate of flow

Hot water cylinder (lagging not shown)

Thermostatic control unit

Gas supply leading to the burner

FIGURE 42   Gas circulator

Flue

Hot water distribution pipe to the taps

Cold water feed pipe from mains or the cold water storage cistern

Circulation of water in the hot water cylinder as it gradually heats up

Flow pipe - Warm water rising from circulator

Return pipe - Water returning to circulator

Gas circulator (sometimes combined with cylinder)

Gas supply

FIGURE 43   Cistern type gas water storage heater

Vent and expansion pipe

Warning pipe - discharge point needs to be considered if not adjacent to an external wall

Hot water storage combined in the same unit as it's cold water supply cistern. All necessary lagging would be added during manufacture

Integral cold water feed pipe

Gas supply. Heating would be as in Figure 41. A flue, not shown, would be necessary if the unit was not room sealed and on an external wall

Cold water supply pipe

Gas fired water heaters *continued*

Anticipative problems	Considerations

length to the width does not exceed 3:1, or they may be of the balanced type if on an external wall, or connected to a common duct which also provided the air for combustion.

All flues must be built with or lined with acid resisting materials. Some of those under former control may be lined with a flexible, metallic duct but unlined lengths may be limited as shown in figure 47. This is seldom applicable to new buildings.

When the lining is of spigot and socket pattern the socket must always face upwards.

Provide adequate clearance between the flues and any combustible material where the flues pass through timber floors, roofs, etc, See figure 48.

6 Location of the flue terminal

Ensure that terminals on traditional flues are positioned at a point where the air may flow freely. See figure 46.

Note No eaves, parapets or projecting piers, etc, must be near the flue outlet otherwise the build up in pressure in that space due to the air movement and the resistance set up by the obstruction may impede the flow of CO_2 and water vapour from the heater. If these products of combustion cannot get away then the build up of CO_2 and possibly some unburnt gas within the room could be a potential danger to the occupants.

Ensure that any terminal fitted to a traditional flue is not less than 600 mm from any ventilator or openable window. See figure 49(a).

Ensure that where projections cannot be avoided that the position of the outlet will comply with Building Regulations. Some illustrations are given in figure 49(b).

Ensure that balanced flue terminals are not located within 300 mm, measured vertically, of any ventilator or openable window. See figure 40.

Note Balanced flues provide the air for combustion and remove the CO_2 and water vapour resulting from it. Both inlet and outlet are in the same zone so the pressures remain equal.

7 Ventilation for combustion

Provide adequate ventilation to comply with the table M9 in Building Regulations. The clear area of the ventilator will vary with the type and size of heater and the capacity of the room in which it is installed.

Note These air vents are essential but may be the cause of considerable annoyance during the winter because of the cold draughts which could flow from them. This unpleasantness can be reduced by either:

(a) locating the fresh air inlet behind a radiator in order that the incoming air is warmed as it enters the room, or

(b) locating the inlet at a high level and preferably near the heating unit in order that the draught will by-pass the activity area in the room. See figure 50.

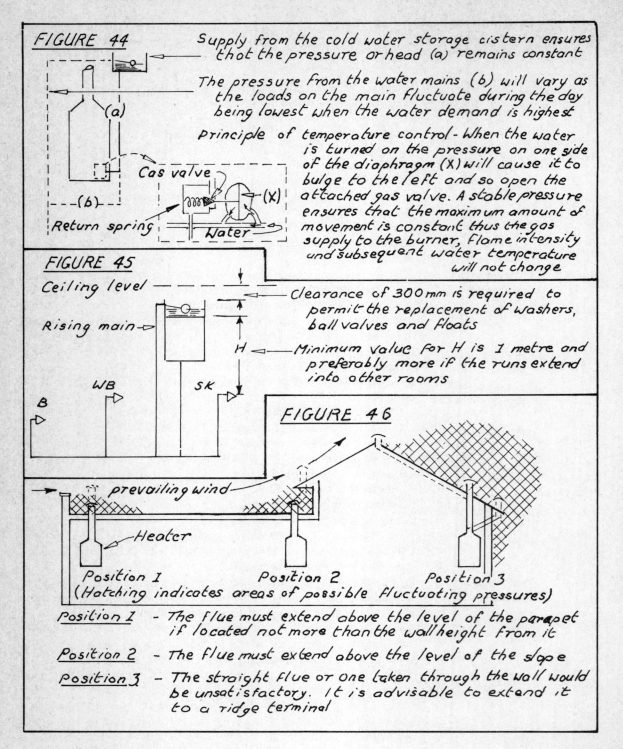

FIGURE 44

Supply from the cold water storage cistern ensures that the pressure or head (a) remains constant

The pressure from the water mains (b) will vary as the loads on the main fluctuate during the day being lowest when the water demand is highest

Principle of temperature control - When the water is turned on the pressure on one side of the diaphragm (X) will cause it to bulge to the left and so open the attached gas valve. A stable pressure ensures that the maximum amount of movement is constant thus the gas supply to the burner, flame intensity and subsequent water temperature will not change

Cas valve

(a)

(b)

Return spring

Water

(X)

FIGURE 45

Ceiling level

Rising main

Clearance of 300mm is required to permit the replacement of washers, ball valves and floats

Minimum value for H is 1 metre and preferably more if the runs extend into other rooms

H

WB

SK

B

FIGURE 46

prevailing wind

Heater

Position 1   Position 2   Position 3
(Hatching indicates areas of possible fluctuating pressures)

Position 1 - The flue must extend above the level of the parapet if located not more than the wall height from it

Position 2 - The flue must extend above the level of the slope

Position 3 - The straight flue or one taken through the wall would be unsatisfactory. It is advisable to extend it to a ridge terminal

Gas fired water heaters *continued*

Anticipative problems	Considerations
8 **Economy in the use of gas**	Ensure that multi-point instantaneous heaters are fitted nearest the tap most frequently used.
	When using gas circulators provide an economy valve to divert the flow as required to control the amount of water being heated. See figure 51.
	See *Hot water systems with central boilers*, Anticipated problems **7** and **9** pages 51 and 53. The latter will not conserve gas when using instantaneous type water heaters but will reduce costs when using storage heaters. See figure 67.

Electric water heaters

The heating element in this case is a copper conductor surrounded with electrical insulation and the whole sheathed in a copper tube. The models available may be instantaneous for short intermittent usage, and storage types for the larger heaters serving several appliances. The element, or immersion heater may be fitted in various positions in the unit and its location can affect the volume and temperature of the hot water drawn off at any one time. This is important for large cylinders with varying demands and if part of a combined system.

When factory made heaters are chosen the element is already installed so the selected model must satisfy the particular needs, but when one is considering a hot water storage cylinder forming part of circulatory systems using an alternative fuel then a choice of positions for the heater is possible, as shown later.

Immersion heaters provide a convenient alternative heat source for combined heating and hot water systems, especially those using any solid fuel because they may still provide heat needed for hot water in summer when the boiler is used for that and space heating is not used.

Factory-made electric water heaters

Anticipative problems	Considerations
1 **Choice of heater to be used**	**(a) Instantaneous heater** As with gas heaters the water is raised to the required temperature as it flows through the unit, so no storage is provided. These are usually single point models serving a sink, wash basin or shower head. Main's models are available but if it is pressure operated then the head needs to be constant. In the case of showers the temperature at the outlet is controlled by the use of a mixing valve. See figure 52. The heater may be fitted above or below the level of the appliance.
	(b) Non-pressure heater These are small hot water storage units in which the water heats up between draw-offs. The tap for controlling the flow is on the inlet side of the unit so that it is not subjected to any

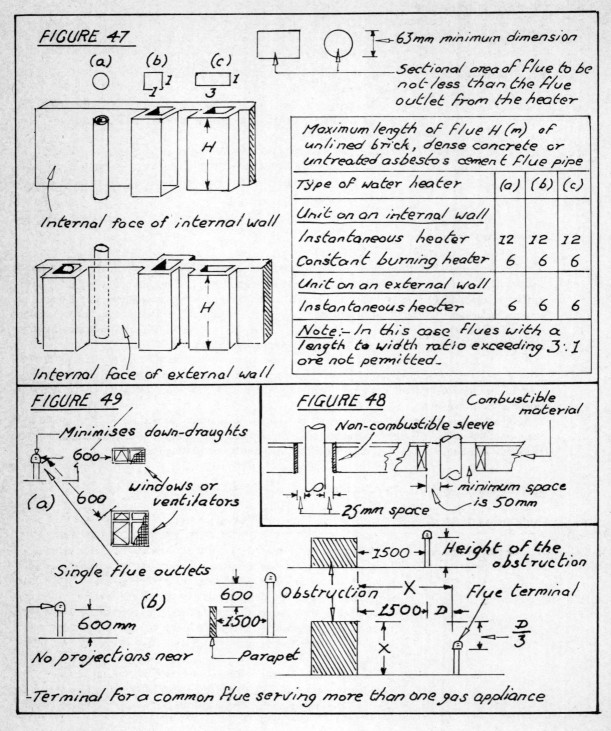

FIGURE 47

(a) (b) (c)

63mm minimum dimension

Sectional area of flue to be not less than the flue outlet from the heater

Internal face of internal wall

H

Internal face of external wall

Maximum length of flue H (m) of unlined brick, dense concrete or untreated asbestos cement flue pipe			
Type of water heater	(a)	(b)	(c)
Unit on an internal wall			
Instantaneous heater	12	12	12
Constant burning heater	6	6	6
Unit on an external wall			
Instantaneous heater	6	6	6

Note:- In this case flues with a length to width ratio exceeding 3:1 are not permitted.

FIGURE 49

Minimises down-draughts

600

600

windows or ventilators

(a)

Single flue outlets

(b)

600

600mm

No projections near

1500

Parapet

Terminal for a common flue serving more than one gas appliance

FIGURE 48

Non-combustible sleeve

Combustible material

25mm space

minimum space is 50mm

1500

Obstruction

X

1500 D

X

Height of the obstruction

Flue terminal

$\frac{D}{3}$

FIGURE 50

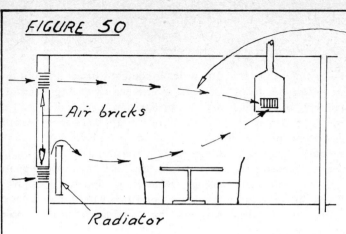

Air bricks

Radiator

High level ventilation opening permits the air currents to by-pass the activity areas if on one wall

If the air vent is located at low level it needs to be on the same wall as the heater or placed behind a radiator which will warm the air in cold weather and reduce the discomfort from cold draughts otherwise possible

Room or space capacity (m³)	Type of water heater	Minimum unobstructed area of vent. (mm²)
Exceeding 6 but not exceeding 11	Instantaneous water heater	3250
Exceeding 6 but not exceeding 11	Any other water heating appliance	9500
Exceeding 11 but not exceeding 21	,,	3250

FIGURE 51

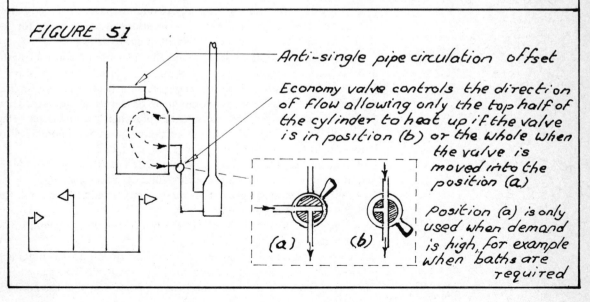

Anti-single pipe circulation offset

Economy valve controls the direction of flow allowing only the top half of the cylinder to heat up if the valve is in position (b) or the whole when the valve is moved into the position (a)

Position (a) is only used when demand is high, for example when baths are required

(a)   (b)

Factory-made electric water heaters *continued*

Anticipative problems	Considerations

pressure or head other than the water it contains, hence its name non-pressure heater. See figure 53. Some models have swivel spouts so that two adjacent wash-basins may be served from a single heater. They must be wall mounted above the level of the taps being served in order that the water may flow by gravity to the appliance being used.

Note The unit is never full and as the water heats up it will expand into the void above and overflow down the inlet pipe causing continuous dripping and water wastage which is prohibited by the Water Byelaws. All non-pressure heaters must be fitted with non-drip devices which, together with the necessary element and lagging is all incorporated during the manufacture.

(c) Pressure type heaters These units fill up completely and as the draw-off tap is on the outlet side of the unit both the tap and the cylinder are subject to pressure either from the mains or from a high level feed cistern. See figure 54. Small heaters are usually fixed to the wall above the appliance but those with a capacity in excess of 100 litres may be floor mounted. When fitting:

1 Provide a cold water storage cistern and feed pipe to replenish the hot water as it is drawn off. If several units are installed on different floors provide a common feed pipe with tap-off points at high level, and and isolating valve for each heater. See figure 56.

2 Provide a vent and expansion pipe per unit if several heaters are required, as a common pipe for this purpose cannot be used.
See figures 54 and 56.

3 Provide a vertical duct for floor mounted units installed below working surfaces to avoid forming perforations in that surface for the rising main and the other pipes just mentioned.

4 Provide a clear space in excess of the element and control length, in front of the heater to ensure that it can be replaced when necessary.

(d) Storage heaters These are large pressure heaters serving several appliances and usually floor mounted or on a special support. They may have one or two heaters dependant upon their use, the latter being an alternative to a gas unit with an economy valve. See figure 55. Provisions are as for pressure units. If they are in cupboards:

1 ensure that the door openings are wide enough to permit the replacement of the whole unit if ever it should develop a leak and

2 provide enough space in front of the cylinder to permit the replacement of faulty elements;

3 provide 500 mm offset near the top of the unit to avoid single pipe circulation. See figure 67.

(e) Cistern type storage heaters These are similar to gas units and must be mounted at high level in order to obtain the necessary head. They have the advantage over the pressure heaters in that the need for

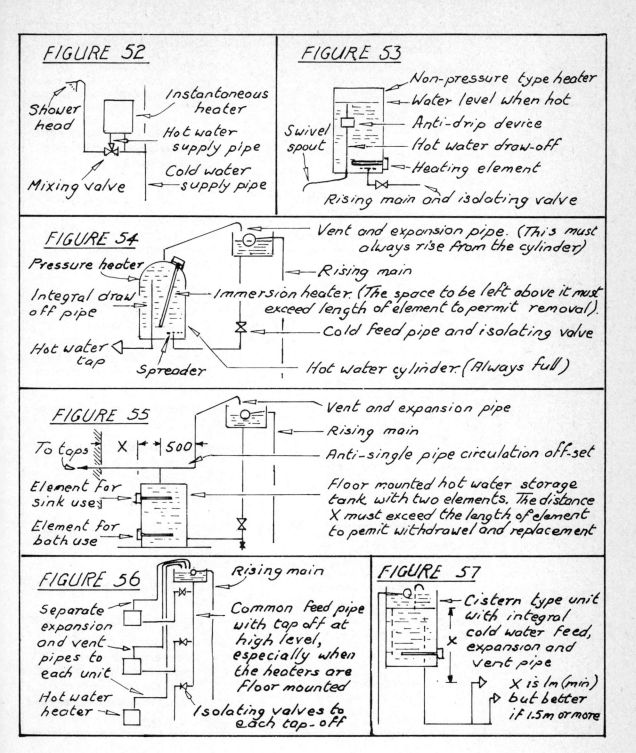

FIGURE 52

Shower head

Instantaneous heater

Hot water supply pipe

Cold water supply pipe

Mixing valve

FIGURE 53

Non-pressure type heater

Water level when hot

Anti-drip device

Hot water draw-off

Heating element

Swivel spout

Rising main and isolating valve

FIGURE 54

Pressure heater

Integral draw off pipe

Hot water tap

Spreader

Vent and expansion pipe. (This must always rise from the cylinder)

Rising main

Immersion heater. (The space to be left above it must exceed length of element to permit removal).

Cold feed pipe and isolating valve

Hot water cylinder. (Always full)

FIGURE 55

To taps

X   500

Element for sink use

Element for bath use

Vent and expansion pipe

Rising main

Anti-single pipe circulation off-set

Floor mounted hot water storage tank with two elements. The distance X must exceed the length of element to permit withdrawel and replacement

FIGURE 56

Rising main

Separate expansion and vent pipes to each unit

Common feed pipe with tap off at high level, especially when the heaters are floor mounted

Hot water heater

Isolating valves to each tap-off

FIGURE 57

Cistern type unit with integral cold water feed, expansion and vent pipe

X is 1m (min) but better if 1.5m or more

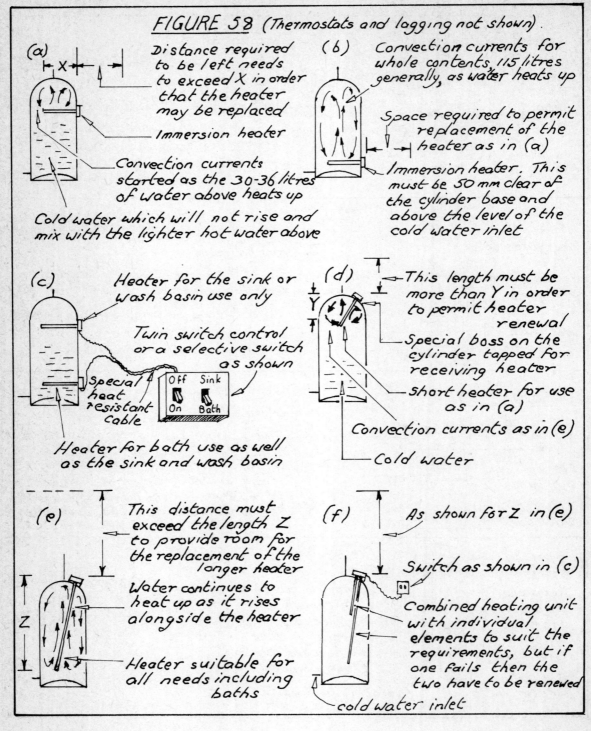

FIGURE 58 (Thermostats and lagging not shown).

(a) Distance required to be left needs to exceed X in order that the heater may be replaced

— Immersion heater

Convection currents started as the 30-36 litres of water above heats up

Cold water which will not rise and mix with the lighter hot water above

(b) Convection currents for whole contents, 115 litres generally, as water heats up

Space required to permit replacement of the heater as in (a)

Immersion heater. This must be 50 mm clear of the cylinder base and above the level of the cold water inlet

(c) Heater for the sink or wash basin use only

Twin switch control or a selective switch as shown

Off   Sink
On   Bath

Special heat resistant cable

Heater for bath use as well as the sink and wash basin

(d) This length must be more than Y in order to permit heater renewal

Special boss on the cylinder tapped for receiving heater

Short heater for use as in (a)

Convection currents as in (e)

Cold water

(e) This distance must exceed the length Z to provide room for the replacement of the longer heater

Water continues to heat up as it rises alongside the heater

Heater suitable for all needs including baths

(f) As shown for Z in (e)

Switch as shown in (c)

Combined heating unit with individual elements to suit the requirements, but if one fails then the two have to be renewed

cold water inlet

44

Factory-made electric water heaters *continued*

Anticipative problems	Considerations

individual vent pipes taken up to the level of the common feed cistern is not required as each has its own built in. See figure 57. Also

1 provide space for the withdrawal of the element should it fail, and the whole unit should it leak;

2 provide a clearance of at least 300 mm above the edge of the cold water cistern to permit the replacement of the ballvalve or parts if needed;

3 provide a warning pipe and determine the position of the actual discharge point.

2 Location of the elements in the hot water cylinders

This is applicable to independant hot water cylinders fitted in circulation systems where the element provides an alternative heat source for use when the boiler fails, or is stopped from operating due to seasonal use. The element or immersion heater may be placed:

1 near the top of the cylinder but on the side as in figure 58(a). This will supply from 30 to 36 litres of hot water which is enough for a wash basin or small sink but inadequate for baths.

Note Only the water above the heater will warm up, decreasing in density as the temperature increases and it circulates. The heavier cold water from below will not mix with it;

2 near the bottom of the cylinder but still on the side as in figure 58(b). All of the water above the heater will now warm up so there will be no shortage of hot water for baths. However, if the occupants are out at work all day it may be uneconomic to keep a cylinder full of hot water if it is not needed as this could result in an unnecessary expense. An alternative is to control the heating periods by the use of a time switch which will come into operation say two hours before the water is required and then switch off once the demand is known to be over;

3 one at the top and the other near the bottom as in 1 and 2 just mentioned. These need to be controlled by individual switches or by the use of a single one having a selector mechanism in order that the correct element is operating depending upon the volume of water required. The result is the same as may be anticipated when using an economy valve with a gas circulator. See figure 58(c);

4 in the cylinder but inserted from the top. There are three variations of this these being

(i) the use of a short element as shown in the figure 58(d) with results as described in 1 but taking less time to reach the required temperature because the convected water will continue to be heated as it rises up alongside the element;

(ii) the use of a long element as shown in the figure 58(e) with results as in 2 together with the accelerated heating rate mentioned in 4(a);

45

Factory-made electric water heaters   *continued*

Anticipative problems	Considerations

(iii) the use of a twin heater as shown in the figure 58(f) making the unit into a two in one heater as described in **3** with the advantage that the cylinder needs only one perforation and boss. It also has the disadvantage in that if one of the elements fails then, as it is a complete unit, both have to be replaced.

Note   When installing hot water cylinders which are fitted with immersion heaters ensure that:

1 the space above or in front of the unit is large enough to permit the removal and replacement of the element should it fail.

2 the 500 mm offset has been provided just above the top of the cylinder to combat any single pipe circulation. See figure 67.

3 lagging is provided if not an integral part of the unit. Sectional panels have an advantage in that part may be removed easily to provide airing for clothes or a necessary heat leak when combined in solid fuel installations.

3   Heat control for the water drawn off

Provide a thermostatically controlled heater to ensure that the temperature of the stored water does not exceed 60°C (140°F).

Note   In hard water areas with no automatic control of the temperature then:

(a) fur will be precipitated from the water and will build up on the actual immersion heater. It will act as an insulator and prevent heat from passing the element to the surrounding water causing the electrical conductors to overheat and eventually burn out. The precipitation of calcium carbonate from the hard water accellerates considerably once the water temperature exceeds 60°C;

(b) the expansion and contraction of the copper cylinder may cause any fur which has been deposited on the sides to drop off and fall to the bottom. If the deposits build up and cover the lower element it will prevent the heat transfer, restrict circulation and eventually the element will overheat and burn out as before.

4   Single pipe and reverse circulation

See *Direct systems for heating water*, Anticipative Problems **8** and **9**.

Hot water systems with central boilers

A boiler containing water is fired by gas, oil or solid fuel and is connected to the hot water cylinder by two tubes called the flow and return pipes. See figure 59.

As the fuel burns so the water heats up and decreases in density. The cooler, heavier water in the return pipe falls and forces the lighter warm water up the flow pipe. As the cool replacement water enters the boiler it in turn warms up and the process is repeated continuously, resulting in what is called primary circulation. The movement of the water is silent and the speed at which it circulates is dependant on

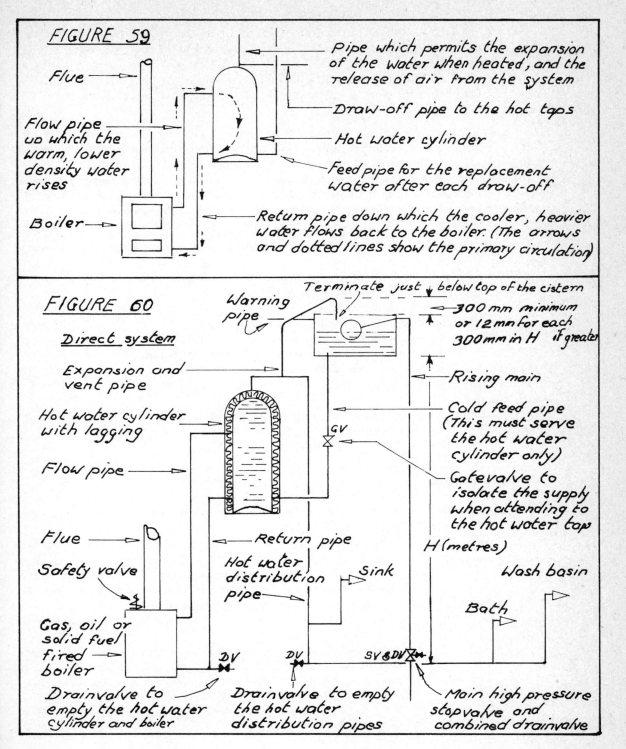

FIGURE 59

Flue

Flow pipe up which the warm, lower density water rises

Boiler →

Pipe which permits the expansion of the water when heated, and the release of air from the system

Draw-off pipe to the hot taps

Hot water cylinder

Feed pipe for the replacement water after each draw-off

Return pipe down which the cooler, heavier water flows back to the boiler. (The arrows and dotted lines show the primary circulation)

FIGURE 60

Direct system

Warning pipe

Terminate just below top of the cistern

300 mm minimum or 12 mm for each 300 mm in H if greater

Expansion and vent pipe

Hot water cylinder with lagging

Flow pipe

Rising main

GV

Cold feed pipe (This must serve the hot water cylinder only)

Gatevalve to isolate the supply when attending to the hot water tap

Flue

Safety valve

Gas, oil or solid fuel fired boiler

Return pipe

Hot water distribution pipe

Sink

Wash basin

Bath

H (metres)

DV

DV

SV & DV

Drainvalve to empty the hot water cylinder and boiler

Drainvalve to empty the hot water distribution pipes

Main high pressure stopvalve and combined drainvalve

Hot water systems with central boilers   *continued*

(a) the difference in the temperature of the water columns in the two pipes and the subsequent densities of them.

(b) the frictional resistance to any flow set up by the walls of the pipes and the bends. The smaller the diameter of the pipes is, and the more bends the water has to negotiate the greater the resistance to flow becomes.

The system used may be one of two categories, these being

(a) *a direct system*   In this all water drawn from the hot water taps has actually passed through the boiler and has been part of the primary circulation.

(b) *an indirect system*   The hot water drawn off from this system has been raised in temperature by heat transferred from the water in the primary circulation pipes to the stored water by the use of a heat exchanger coil within the cylinder, the whole unit being a calorifier.

Direct systems for heating water See figure 60. This will comprise

1 a floor mounted boiler heated by gas, oil or solid fuel or in some cases a gas fired unit hung on a wall;

2 a hot water storage cylinder of not less than 115 litres capacity and this will always be full;

3 connecting tubes termed the flow and return pipes;

4 provisions common to other installations such as:

(a) a cold feed pipe to replenish the system with fresh water after each draw-off. This must be connected near the base of the hot water cylinder.

(b) a combined expansion and vent pipe to permit the water which is being heated to expand without spilling over into areas where it is not wanted and for air which may get into pipes to escape harmlessly. It must rise from the top of the cylinder and discharge above the cold water supply cistern.

(c) distribution pipes to carry hot water from storage to the hot water taps together with the necessary isolating and drain valves.

Anticipative problems	Considerations
1 The possible formation of fur or scale in pipes	Never use this system in hard water areas or when the installation also includes space heating. *Note* The rain absorbs CO_2 from the air as it falls, and calcium carbonate as it seeps down through the earth if the ground is of chalk or limestone, thus making the water hard. When this water is heated the CO_2 is driven off and the calcium carbonate is precipitated as fur, the process accelerating as the temperature rises above $60°C$. This fur is rock hard and builds up on the pipe walls reducing the bore and slowing up the flow. See figure 61(a). Because circulation is restricted the water will stay in the boiler longer than it would otherwise, and it may boil thus forming steam. If this cannot escape then the build up of pressure resulting from it could cause an explosion. Whilst safety valves may be installed in order to avoid this, they too can become furred up and inoperative unless periodically inspected and serviced.

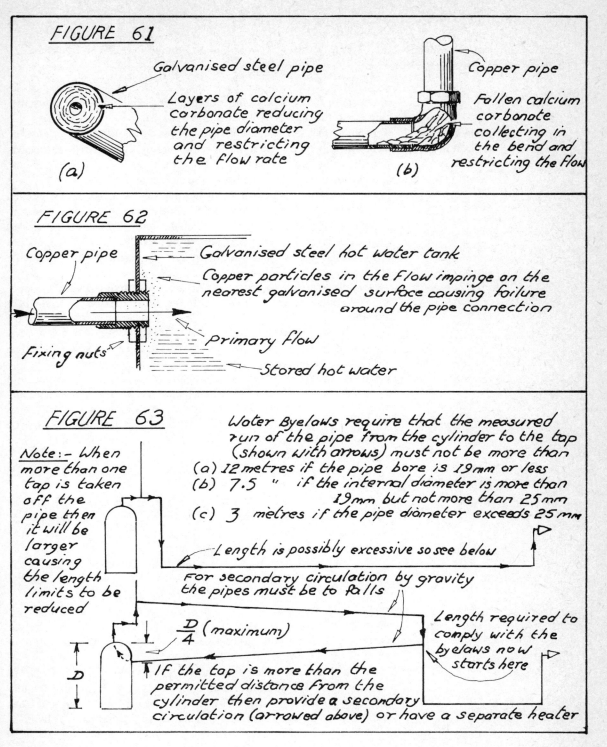

FIGURE 61

Galvanised steel pipe

Layers of calcium carbonate reducing the pipe diameter and restricting the flow rate

(a)

Copper pipe

Fallen calcium carbonate collecting in the bend and restricting the flow

(b)

FIGURE 62

Copper pipe

Galvanised steel hot water tank

Copper particles in the flow impinge on the nearest galvanised surface causing failure around the pipe connection

Primary flow

Fixing nuts

Stored hot water

FIGURE 63

Note:- When more than one tap is taken off the pipe then it will be larger causing the length limits to be reduced

Water Byelaws require that the measured run of the pipe from the cylinder to the tap (shown with arrows) must not be more than

(a) 12 metres if the pipe bore is 19mm or less

(b) 7.5 " if the internal diameter is more than 19mm but not more than 25mm

(c) 3 metres if the pipe diameter exceeds 25mm

Length is possibly excessive so see below

For secondary circulation by gravity the pipes must be to falls

$\frac{D}{4}$ (maximum)

Length required to comply with the byelaws now starts here

D

If the tap is more than the permitted distance from the cylinder then provide a secondary circulation (arrowed above) or have a separate heater

Hot water systems with central boilers *continued*

Anticipative problems	Considerations

With copper pipes the expansion and contraction of the metal causes the scale to flake off and fall, collect in bends and junctions and eventually block the pipe as before. See figure 61(b).

2 Expansion of the water as it heats up

Provide a vent and expansion pipe from the top of the hot water cylinder. The pipe must extend above the supply cistern for one twenty-fourth of the head in metres, and not less than 300 mm, See figure 60, and terminate over but just below the cistern top.

Note This may pose a problem in buildings with flat or low pitched roofs unless special tank rooms can be constructed at high level to accommodate it. Alternatively the cistern could be on the top floor provided that there were not hot water taps to be served on this level, or should this be unavoidable, that the cistern base is at least a metre above the level of the tap.

3 Electrolysis

Never mix metals in hot water systems, for example use only copper pipes with a copper cylinder. Also use bowerbarffed boilers where possible.

Note If water flows from copper towards galvanised steel tiny particles of copper may be carried along with it and impinge on the galvanising. Electrolitic action will occur and the protective zinc in the galvanising will be destroyed followed by the corrosion of the steel and eventually leaks will develop. When mixed metals are used in conjunction with heat and circulating water the deterioration rate is accelerated. See figure 62. If water flows from the galvanised steel to the copper then no action will occur, unless the flow path should form a ring.

4 Selection of the heat source and ancilleries

Provide a heat source and any other units which are necessary for its satisfactory functioning. This may be;

(a) a wall or floor mounted gas boiler, and a flue which may be either a traditional one terminating above roof level or a balanced type fitted direct on the external wall. See figure 40. Existing flues would have to be lined with a flexible metal tube;

(b) an oil-fired boiler, flue and fuel tank. The flue is usually of the traditional type but existing ones need to be fitted with a flexible metallic liner. The fuel store needs to be located at a higher level than the boiler in order to provide a natural gravity flow from it wherever possible, otherwise it needs to be pumped. Generally it will be sited externally and must be easily accessible for the tanker hose unless an extended filling point and pipe line is provided;

(c) a solid fuel boiler, flue and fuel store. (See section on *Central Heating*). The flue is traditional and must comply with Building Regulations. The fuel store needs to be easily accessible for the consumer and the delivery man, and preferably located as an integral part of the main

Hot water systems with central boilers *continued*

Anticipative problems	Considerations

	building. Passageways leading to it must be wide enough to permit the coalman to turn with a sack on his shoulder without dirtying or damaging the walls, especially at the fuel discharge point.
	Note Ensure that there is adequate space at the sides and in front of any boiler in order that the cover may be removed for servicing it. Some cover fixings are at the rear and so space will be needed on either side to permit the service engineer to reach back and release them.
5 Water supply for the system	Provide a cold water cistern to replenish the system after each draw-off from the hot taps, if combined hot and cold water storage is not already allowed for. It must be above the level of the hot water cylinder it feeds, be not less in actual capacity, be controlled at the inlet with a ballvalve and fitted with its own warning pipe.
6 Location of the hot water cylinder	Provide a hot water storage cylinder of the same metal as used for the the pipes and site it above level of the boiler if reverse circulation is anticipated, but below the level of the feed cistern in order that no draw-off pipe ever goes above this cistern otherwise no draw-off will be possible from some hot taps.
	Note The highest water level in the system when it is cold will never be higher than that in the feed cistern when vented to the atmosphere. See figure 64.
7 Distribution pipe runs	1 Ensure that the distance measured from the cylinder to the taps along the actual run of the pipe complies with Water Byelaws. The maximum permissible length will vary with the bore of the pipe used. See figure 63.
	Note If the lengths are excessive they are called 'deadlegs'. When the outlet tap is closed the pipe is left full of hot water, which, if not drawn off soon will cool and the next user too often tends to let this run to waste before inserting the plug and collecting his hot water. Should this now be too hot then more cold may be added. A water regulation is drawn up to reduce the amount of water which may be left in the pipes and later run to waste.
	2 Provide freedom for the pipes to expand when bedded in solid floors, especially at bends and tees where the restriction may be greatest. See figure 65.
	3 Provide low pressure gatevalves on the cold feed pipe to the cylinder and on branch pipes which are serving several appliances.
	Note These valves are necessary to isolate the supply during maintenance work to the hot water taps, and must be easily accessible. See figure 60.
	4 Provide drain valves to each drop and to the boiler in order that the system may be emptied. These also must be easily accessible. See figure 60.

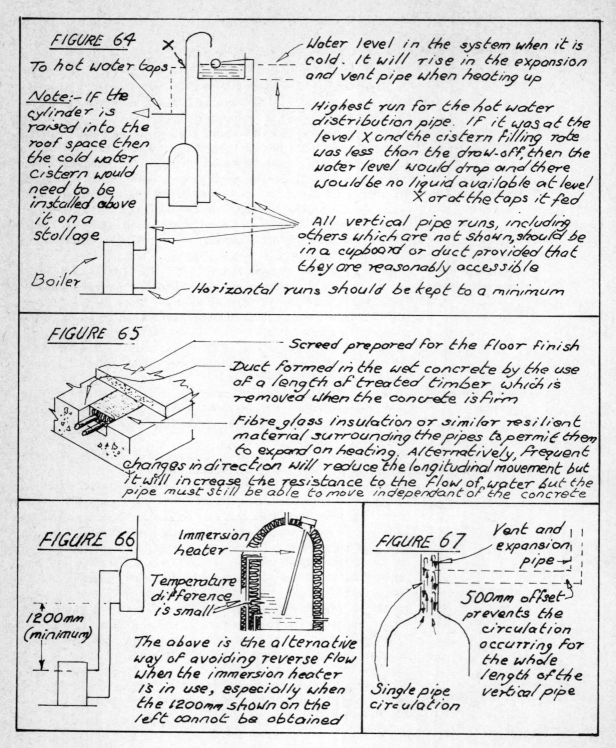

FIGURE 64

To hot water taps

Note:- If the cylinder is raised into the roof space then the cold water cistern would need to be installed above it on a stollage

Boiler

Water level in the system when it is cold. It will rise in the expansion and vent pipe when heating up

Highest run for the hot water distribution pipe. IF it was at the level X and the cistern filling rate was less than the draw-off, then the water level would drop and there would be no liquid available at level X or at the taps it fed

All vertical pipe runs, including others which are not shown, should be in a cupboard or duct provided that they are reasonably accessible

Horizontal runs should be kept to a minimum

FIGURE 65

Screed prepared for the floor finish

Duct formed in the wet concrete by the use of a length of treated timber which is removed when the concrete is firm

Fibre glass insulation or similar resilient material surrounding the pipes to permit them to expand on heating. Alternatively, frequent changes in direction will reduce the longitudinal movement but it will increase the resistance to the flow of water but the pipe must still be able to move independant of the concrete

FIGURE 66

Immersion heater

Temperature difference is small

1200mm (minimum)

The above is the alternative way of avoiding reverse flow when the immersion heater is in use, especially when the 1200mm shown on the left cannot be obtained

FIGURE 67

Vent and expansion pipe

500mm offset prevents the circulation occurring for the whole length of the vertical pipe

Single pipe circulation

Hot water systems with central boilers *continued*

Anticipative problems	Considerations
8 **Reverse circulation**	If an immersion heater is to be installed as well as a boiler then ensure that the bottom of the cylinder is not less than 1200 mm above the level of the flow outlet on the boiler, or keep the flow and return pipes relatively close and if possible enclose them within the thickness of the cylinder lagging. See figure 66. *Note* Reverse circulation may occur when the immersion heater is on and heated water may flow from the cylinder down the flow pipe and along to the boiler before returning to the cylinder through the return pipe. This circulation if left unchecked can lead to considerable wastage of electricity and a great deal of unnecessary expense.
9 **Single pipe circulation**	Provide a 500 mm offset on the vent and expansion pipe immediately above the top of the hot water storage cylinder. See figure 67. *Note* Without the offset mentioned above the hot water will rise up the centre of the vent and expansion pipe and then be cooled by the walls of the pipe causing it to fall and start circulating within that single pipe. This again results in a wastage of electricity and unnecessary high costs.
10 **Freezing**	Ensure that all hot water pipes in the roof and ventilated spaces are lagged.

The indirect system for water heating *(See figure 68)*

This varies from the direct system in that it has:

   1   a second feed cistern designed for the primary water only, and this has its own feed pipe and vent and expansion pipe;

   2   a calorifier or indirect cylinder which has a small tank or coil within it, through which the primary water flows transferring its heat to the stored draw-off water around it whilst at the same time remaining separated from it.

Indirect systems should always be used

   (a)   in hard water areas. The primary water is not changed each time the hot taps are used, as in direct systems, therefore after the deposit of fur from the initial fill, little or no further precipitation can occur.

   (b)   when heating and hot water systems are combined, any draw-off of water from the hot taps from a combined direct heating and hot water system will cause a loss of heat to the radiator circuit as the cold replacement water flows in. If the hot water is drawn for a bath then the radiator system could run cold. This cannot occur in indirect systems as the primary and draw-off water never mix.

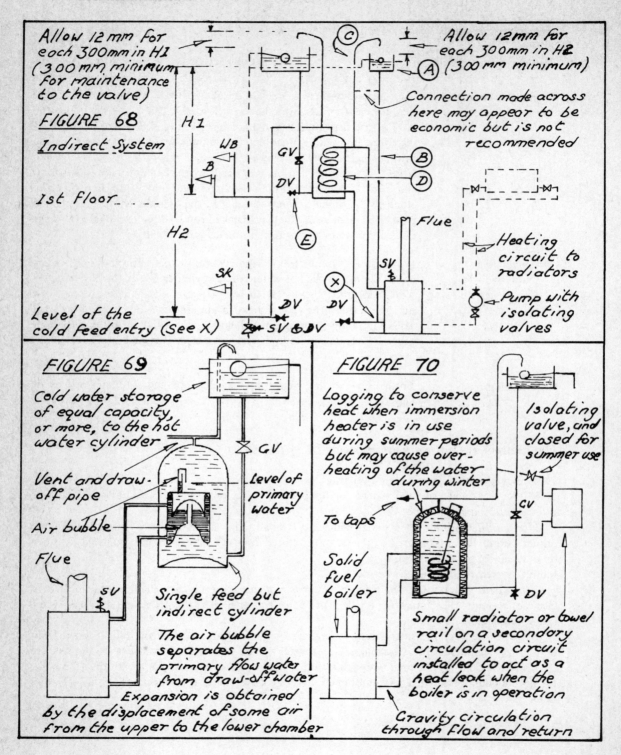

Allow 12mm for each 300mm in H1 (300mm minimum for maintenance to the valve)

FIGURE 68

Indirect System

1st floor

Level of the cold feed entry (See X)

H1

H2

WB

B

GV

DV

E

SK

DV

SV & DV

DV

X

SV

Flue

(C)

(A)

(B)

(D)

Allow 12mm for each 300mm in H2 (300 mm minimum)

Connection made across here may appear to be economic but is not recommended

Heating circuit to radiators

Pump with isolating valves

FIGURE 69

Cold water storage of equal capacity, or more, to the hot water cylinder

Vent and draw-off pipe

Air bubble

Flue

SV

GV

Level of primary water

Single feed but indirect cylinder

The air bubble separates the primary flow water from draw-off water Expansion is obtained by the displacement of some air from the upper to the lower chamber

FIGURE 70

Logging to conserve heat when immersion heater is in use during summer periods but may cause over-heating of the water during winter

To taps

Solid fuel boiler

Isolating valve, and closed for summer use

CV

DV

Small radiator or towel rail on a secondary circulation circuit installed to act as a heat leak when the boiler is in operation

Gravity circulation through flow and return

Anticipative problems	Considerations
1 Pipes and the fittings needed in addition to those for direct systems	**1** Provide a small feed cistern of a nominal capacity of not less than 45 litres, together with its own ballvalve and warning pipe for the primary circulation circuit. It really acts as a topping up tank for the primary water should any losses occur resulting from evaporation. See A in figure 68.
	2 Provide a separate feed pipe connecting this cistern to the heating return near the boiler.
	Note Connecting the feed pipe to the vent and expansion pipe saves a lot of tubing but is not recommended as other problems can arise due to the cold water trying to fall under the force of gravity, and the hot water trying to rise up in the same pipe as a result of being heated. See B in figure 68.
	3 Provide an indirect cylinder for the heating and storage of hot water. See D in figure 68.
	4 Provide a drainvalve at the lowest part of the feed pipe to the indirect cylinder to ensure that it can be emptied. See E in figure 68.
	5 Provide a separate vent and expansion pipe for the primary water, and terminate it over its own feed cistern. See C in figure 68.
	6 Determine the discharge point for the warning pipe mentioned in **1**. See figures 32 to 37.
2 Furring up of pipes in hard water areas	This should not occur on the primary side of the installation but the temperature of the draw-off water must be controlled by a thermostat and not allowed to rise above 60°C (140°F). The reason for the non-furring of the primary water is because it is not changed, and after the initial deposit when the pipes are first filled any further precipitation is negligible.
3 Corrosion of the pipes	This is usually the result of oxidation and will be negligible because if the water is not changed then no fresh oxygen will be drawn into the system.
4 Loss of hot water	This may occur due to the ballvalve in the feed cistern for the primary water seizing up so that any water which has evaporated is not replaced. Periodic checks on this cistern are advisable, or the provision of a diaphragm-type ballvalve may be advantageous in that the working mechanism is always dry.
5 Direct system's problems	See **2**, **3**, **4**, **5**, **6**, **7**, **8**, **9** and **10** as referred to in 'direct systems'.

Direct systems acting as indirect installations

Hot water cylinders are available where the installation is direct in appearance as it only has one feed cistern, but it acts as an indirect system in that the primary and draw-off waters are separated by an air bubble. See figure 69. Whilst these work satisfactorily by gravity or pumped circulation, care is needed in the selection of the pump to ensure that any pressures which may build up will not be such as to dislodge this pocket of air. If it is lost it will reform again but it may take several hours. Any expansion of the primary water on heating is taken up by the compression of the air and partial displacement of the bubble into the lower chamber.

Problems applicable to combined systems

This concerns those installations where the water is heated by the use of solid fuel in the winter, and electricity, using an immersion heater, during the summer or such times when the boiler is out of action.

Anticipative problems	Considerations
1 **Continuity of heat dispersal from the solid fuel**	Provide gravity circulation for the domestic hot water heating section of the installation. *Note* In cases of overheating the fire can be terminated automatically if the fuel is gas, oil or electricity, but with solid fuel it cannot be put out immediately and the cooling may take some considerable time. Gravity circulation ensures that any excess heat can be dissipated safely and usefully whilst the cooling occurs.
2 **Alternative heat source for summer use**	1 Provide an immerison heater for use in the summer when the actual space heating is no longer required, and provide a time clock and thermostat to ensure that it only operates when required and the temperature limits are controlled. 2 Provide lagging around the hot water cylinder otherwise the heat loss will be excessive. Sectional lagging has an advantage in that it can be removed easily if required when the solid fuel is functioning again, assuming that heat leaks have not been provided.
3 **Overheating**	Provide a heat leak in the form of a small radiator, rail or airing cupboard coil for use when the heating is again from solid fuel. *Note* It is inconvenient to refit and remove lagging at each change of season, so the heat leak eliminates the need for this, and if it is incorporated in the gravity flow section it will dissipate any excess heat usefully and safely. Provide an isolating valve for this heat leak which must be closed when the immersion heater is operating but open when the solid fuel fire is on. See figure 70.

Hot water by solar energy

Solar energy as a heat source is a debateable factor in this country and several basic factors listed below have to be resolved before it can be accepted as a satisfactory alternative to existing ones.

  1 *Energy availability* In winter when the heat is most needed it is not available in this country.

  2 *Temperature available* The maximum temperature of the water from a solar collector during winter is about 50°C and this is not constant. If the draw-off water has to be 60°C then any solar system can only operate as a preheater.

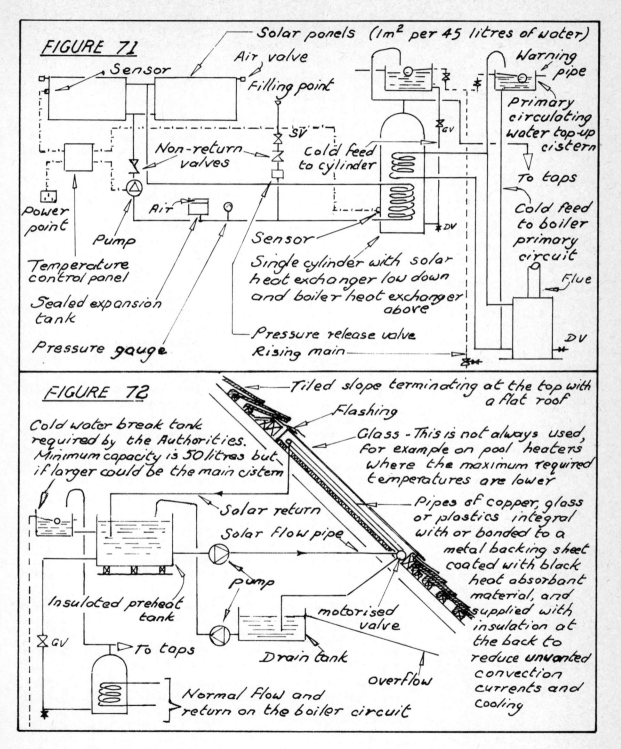

FIGURE 71

Solar panels (1m² per 45 litres of water)

Sensor

Air valve

Filling point

Warning pipe

Primary circulating water top-up cistern

SV

Non-return valves

Cold feed to cylinder

To taps

Cold feed to boiler primary circuit

GV

DV

Power point

Air

Pump

Sensor

Flue

Temperature control panel

Single cylinder with solar heat exchanger low down and boiler heat exchanger above

Sealed expansion tank

DV

Pressure gauge

Pressure release valve

Rising main

FIGURE 72

Cold water break tank required by the Authorities. Minimum capacity is 50 litres but if larger could be the main cistern

Tiled slope terminating at the top with a flat roof

Flashing

Glass - This is not always used, for example on pool heaters where the maximum required temperatures are lower

Solar return

Solar flow pipe

Pipes of copper, glass or plastics integral with or bonded to a metal backing sheet coated with black heat absorbant material, and supplied with insulation at the back to reduce unwanted convection currents and cooling

GV

Insulated preheat tank

Pump

To taps

motorised valve

Drain tank

Normal flow and return on the boiler circuit

overflow

57

3 *Initial costs*   As the solar energy plant can only act as a preheater then a normal heating installation must still be available, which means possibly doubling the initial installation costs plus increased loan interest to be paid. Running costs may be reduced a little in that the normal system will be heating pre-heated water instead of very cold.

4 *Storage of heat when available*   This problem is not yet resolved. If solar energy is to be used then the units required are

    (a)   solar heating panels or collectors;

    (b)   a solar primary circulating circuit, generally pumped;

    (c)   a hot water storage vessel which passes preheated water to a normal storage cylinder, both of them being indirect. Alternatively a dual type cylinder may be used with the solar heating coil being placed below the normal coil used with the boiler. See figure 71;

    (d)   an expansion vessel, as open venting discharging into a cold water cistern at the top of the system is difficult to achieve when the collector panels are located near the ridge on pitched roofs and on flat roofs;

5   a drain tank to collect the water run off prior to a temperature drop which may cause freezing when no other frost precautions have been adopted.

Solar heated hot water systems

Anticipative problems	Considerations
1 The location of solar panels	1 Ensure that the panels are in a non-shaded area and properly oriented, that is generally facing south or within 10° of it and tilted at an angle of 30 to 50 degrees from the horizontal. *Note* Consider adjacent trees and their growth, and the possibility of new buildings or extensions to existing ones, any of which could later provide permanent shade and make the solar system useless. *Note* The location should be accessible in order that the panels may be cleaned periodically. Rain will wash off the dust which may settle but other deposits resulting from the fouling by birds cannot be removed naturally without manual effort.

2 Choice of the material for the primary flow	Copper is satisfactory but can be expensive, or polypropylene if the Authorities approve. Avoid
	    (a) steel as it corrodes in air and water;
	    (b) aluminium which may also corrode in time;
	    (c) polycarbonate if affected by sunlight.

3 Freezing up of pipes in winter	Provide antifreeze for the primary flow but if the system is open it may oxidise. Provide an automatic drain valve and drain tank to take the water once the temperature drops below a certain limit, otherwise the external pipes may freeze. See figure 72.
4 Expansion of the water heated by solar energy	Provide a sealed system of pipes for the primary flow with an expansion vessel to take up the increase in volume as the water heats up. See figure 71. A diaphragm separates the hot water from an air chamber and as the water volume increases the air is compressed and provides the space necessary for expansion.

58

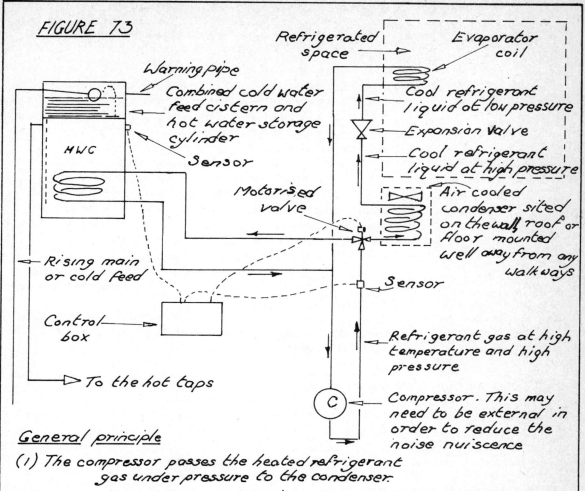

FIGURE 73

Refrigerated space

Evaporator coil

Warning pipe

Combined cold water feed cistern and hot water storage cylinder

Cool refrigerant liquid at low pressure

Expansion valve

Cool refrigerant liquid at high pressure

HWC

Sensor

Motorised valve

Air cooled condenser sited on the wall, roof or floor mounted well away from any walkways

Rising main or cold feed

Sensor

Control box

Refrigerant gas at high temperature and high pressure

To the hot taps

C

Compressor. This may need to be external in order to reduce the noise nuiscence

General principle

(1) The compressor passes the heated refrigerant gas under pressure to the condenser.

(2) The air cooled condenser dissipates the heat to the external air the refrigerant now changing to a liquid, still pressurised

(3) The expansion valve causes a reduction in pressure but still leaves the refrigerant as a cool liquid.

(4) At the evaporator coil the cool refrigerant liquid changes back to a gas using the latent heat from the surrounding air. The warm gas is at low pressure as it flows back to compressor

Note:- If the motorised valve is incorporated the flow may be diverted automatically to a hot water cylinder and the heat can be used to heat the draw off water instead of being passed direct to the atmosphere. When the hot water reaches the required temperature, sensors will activate the control and divert the flow to the condenser as is usual

Solarheated hot water systems *continued*

Anticipative problems	Considerations
	Note Open vents may be provided but they need to be above the level of the panels. This is difficult and even impractical when the collectors are on the roof. Also maintenance on the cistern located so high is difficult. A low level alarm is required to indicate any loss of water due to evaporation or lack of replenishment resulting from a stuck piston in the ballvalve.
5 Provision of draw-off water at the required temperature	Provide a solar hot water cylinder of equal volume to the normal hot water storage vessel, together with its feed pipe, gate valve and drain valve, and where applicable, an expansion and vent pipe. Provide for future replacement and also a pump for the primary water circuit. *Note* The pump will be controlled by sensors on the solar panel and the cylinder, the action being started and stopped according to the temperature differences selected. See figure 71. Alternatively provide a single dual type storage cylinder having first considered the extra height required and the replacement problems. See figure 71.
6 Ancillary problems	Where extra cisterns are provided allow space around it for insulation, height above for the expansion and vent pipe and to permit maintenance, and a ballvalve and warning pipe as before.

Hot water from refrigeration

Water may be warmed by the heat extracted from the cooled areas and the compressor. It operates intermittently over every 24 hours. See figure 73.

Anticipative problems	Considerations
1 As mentioned already	This will cover all points previously covered regarding hot water cylinder, cold water feed and supply cistern, head, pressure, expansion, etc.
2 Power supply	Check whether single or three phase electrical supply is required for the proposed unit.
3 Noise and location of the units	1 Locate the cooling condenser, the fan, and if possible the compressor externally where any noise from it will not provide another problem. 2 Ensure that all mechanical and electrical parts are protected from the weather, and that they are easily accessible for maintenance. 3 Ensure that the condenser is provided with an uninterrupted flow of air. The direction of flow will vary with the type so consult the manufacturer or supplier regarding important requirements.
4 Long refrigerant runs	Avoid long runs in excess of 25 m otherwise install a break tank and a pumped secondary circuit using water as the heat transfer medium. Also provide some antifrost measures for pipes which carry water and have to run externally.

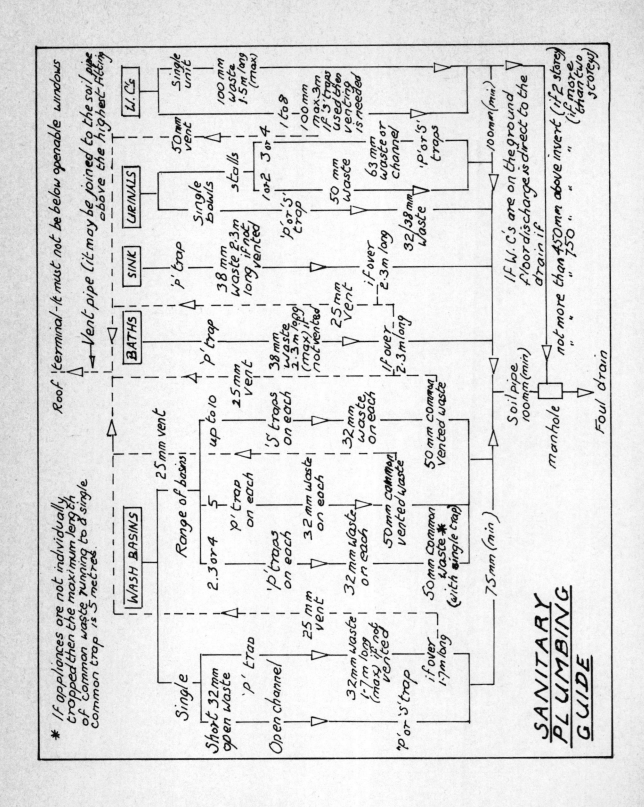

SANITARY PLUMBING GUIDE

3 Sanitary plumbing

Sanitary plumbing is a system of pipes installed to permit the transfer of waste water and sewage from the building to the foul drain. Also it provides a means of ventilation for that drain in order that there can be no build up of unpleasant odours or methane gas within the system which might accidentally permeate into the building.

Generally it is a single stack pipe extending vertically from the drain in a straight line to a point which is above roof level in most cases, and above all openable windows. The appliances are connected to the stack by branch pipes called *wastes*, all of which have a trap filled with water which prevents the smells and gases mentioned above from permeating into the rooms. See figure 74.

Research has established that if the lengths, diameters and falls of the branches exceed certain limits, or the traps be of particular design then pressure variations can occur within the pipes causing the water seal to be drawn out thus making the trap useless. If the rules cannot be conformed to, then a vent pipe must be provided to allow fresh air to enter and balance these pressures, turning the installation into a one pipe or modified one pipe system. See figures 74(b) and (c). This is more expensive because of the second stack and vent pipes which are required. If the appliances can be grouped around a single stack and the limitations on branch lengths, diameters and falls adhered to then the single stack system is simpler and more economic.

Most appliances will discharge direct to the soil stack but in some cases those on the ground floor may be arranged differently. WCs may be taken direct to the drain at the nearest manhole, and waste pipes for ground floor appliances to trapped gulleys. These are the only waste pipes to be permitted externally. The stack pipe may be on an external wall for up to three storeys only, but it is better for the installation and maintenance on branches if kept internal provided space can be found for it, and the clearing eyes and traps made easily accessible.

Anticipative problems	Considerations
1 Loss of water seal due to (a) induced siphonage	1 Provide waste pipes with the lengths, falls, diameters and traps as recommended in CP 304 if single stack plumbing is required. In some cases the lengths may be extended provided that the pipe diameter is increased. See figure 75.
	2 Provide 'P' traps with 75 mm seals, and short straight wastes with swept connections to a 50 mm diameter straight common waste when a range of up to four wash basins is considered but venting is not required. See figure 77(a).
	3 Provide all as in 2 plus a 25 mm common vent connecting to a 50 mm diameter vertical vent pipe if the basin range is extended to five units. See figure 77(b).
	4 Provide anti-siphon pipes to all appliances when the range extends up to ten units and using 'S' traps. In all cases ventilation branches must rise at an angle of not less than $12.5°$. See figure 78.
	Note As a discharge runs down the wall of a soil pipe, or along a common waste pipe, it induces air to flow from the branch connected to it thus reducing the pressure in that branch to less than atmospheric.

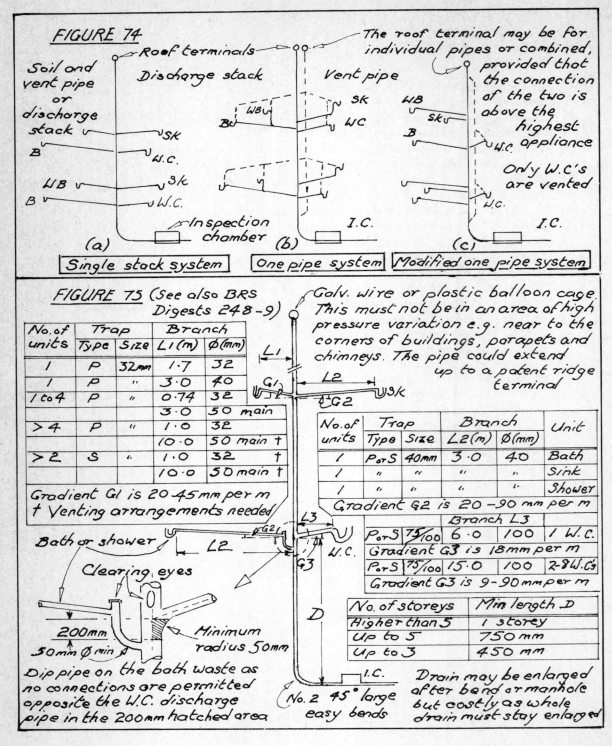

FIGURE 74

Soil and vent pipe or discharge stack

Roof terminals

Discharge stack

Vent pipe

The roof terminal may be for individual pipes or combined, provided that the connection of the two is above the highest appliance

Only W.C's are vented

SK
B
W.C.
WB
B
SK
W.C.
WBU
SK
WC
WB
SK
B
W.C.

Inspection chamber

I.C.

I.C.

(a) Single stack system

(b) One pipe system

(c) Modified one pipe system

FIGURE 75 (See also BRS Digests 248-9)

No. of units	Trap Type	Trap Size	Branch L1(m)	Branch Ø(mm)
1	P	32mm	1.7	32
1	P	"	3.0	40
1 to 4	P	"	0.74	32
			3.0	50 main
>4	P	"	1.0	32
			10.0	50 main †
>2	S	"	1.0	32 †
			10.0	50 main †

Gradient G1 is 20·45mm per m
† Venting arrangements needed

Galv. wire or plastic balloon cage. This must not be in an area of high pressure variation e.g. near to the corners of buildings, parapets and chimneys. The pipe could extend up to a patent ridge terminal

L1
L2
G1
SK
G2

No. of units	Trap Type	Trap Size	Branch L2(m)	Branch Ø(mm)	Unit
1	P or S	40mm	3.0	40	Bath
1	"	"	"	"	Sink
1	"	"	"	"	Shower

Gradient G2 is 20-90 mm per m

			Branch L3		
P or S	75/100	6.0	100	1 W.C.
Gradient G3 is 18mm per m				
P or S	75/100	15.0	100	2-8 W.Cs
Gradient G3 is 9-90mm per m				

Bath or shower
Clearing eyes
L2
L3
G2
W.C.
G3

D

No. of storeys	Min length D
Higher than 5	1 storey
Up to 5	750 mm
Up to 3	450 mm

200mm
50mm Ø min
Minimum radius 50mm

Dip pipe on the bath waste as no connections are permitted opposite the W.C. discharge pipe in the 200mm hatched area

I.C.

No. 2 45° large easy bends

Drain may be enlarged after bend or manhole but costly as whole drain must stay enlarged

Loss of water seal *continued*

Anticipative problems	Considerations
(b) Self siphonage	The air pressure on the appliance side of the seal is now greater than that in the branch and will push the water from the trap into the waste pipe, thus breaking the seal and making it inoperative. See figure 76.

(b) Self siphonage

The air pressure on the appliance side of the seal is now greater than that in the branch and will push the water from the trap into the waste pipe, thus breaking the seal and making it inoperative. See figure 76.

1 Avoid the use of appliances which have no flat or relatively flat bottom. Baths and sinks do not have this problem as the final part of discharge is slow and so replenishes any part of the seal lost during the normal surge.

2 Avoid sharp bends in the waste pipes which have the same diameter as the trap outlet, as this causes the pipe to flow full. The momentum of the flow and the resulting negative pressure behind will draw water from the trap as the flow ceases and make the seal inoperative. If the discharge pipes are larger in diameter than the outlets then the pipe cannot flow full, and air travelling above the flow will eliminate the siphonic action causing the problem.

Note The use of resealing traps should not be considered as the solution to the problem. The traps have small chambers which fill with water during the initial flow and then empty when it ceases thus making up for any water lost from the seal due to that discharge. The scum, hair etc which is normally deposited in the trap will also build up in the channels and chambers eventually filling them and turning the resealing trap into a normal one. It will work successfully again when cleaned out but the maintenance costs must be higher than with the normal flow traps. For the priciple of these traps see figure 80.

(c) back pressure

1 Provide a large radius bend at the base of the stack. This may be accompanied by an increase in the diameter of the drain, but as this would have to be retained for the whole length of the run it would increase the costs considerably.

2 Run WCs located on the ground floors direct to the manhole as recommended in table 3 in the BRS Digest 249 if the vertical measurement taken from the bottom of the WC outlet to the invert of the drain is less than

    (a) 1 storey for buildings over 5 storeys

    (b) 750 mm for buildings up to 5 storeys

    (c) 450 mm for buildings up to 3 storeys.

See figures 75 and 79.

3 Run ground floor wastes direct to the drain through a trapped gulley. These pipes may be taken externally as shown in figure 79. Usually units other than baths and showers are not affected as the wastes are higher off the floor.

Note If the flow rate of a discharge slows up because of the bend at at the base of the stack and is followed quickly by another, then air may be trapped between the two. This compressed air will be forced back up those lower branches and out through the trap causing turbu-

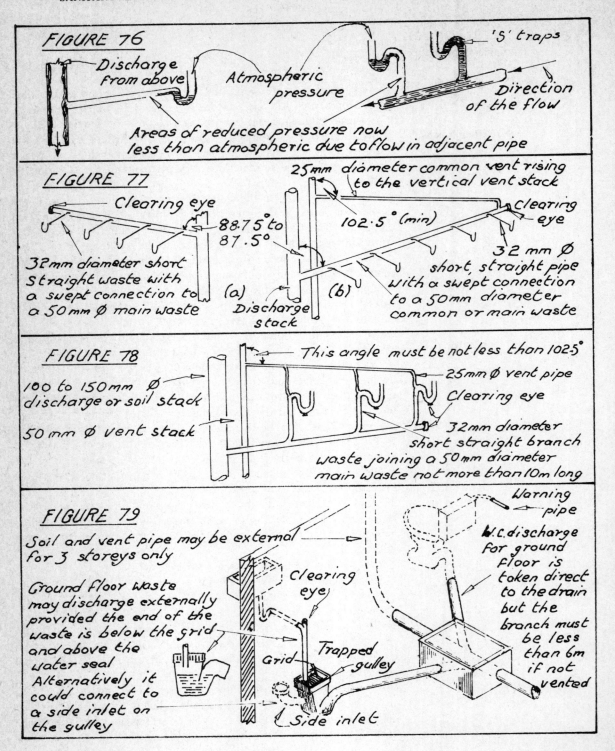

FIGURE 76

Discharge from above

Atmospheric pressure

'S' traps

Direction of the flow

Areas of reduced pressure now less than atmospheric due to flow in adjacent pipe

FIGURE 77

Clearing eye

88.75° to 87.5°

25mm diameter common vent rising to the vertical vent stack

102.5° (min)

Clearing eye

32mm diameter short straight waste with a swept connection to a 50mm Ø main waste

(a)

Discharge stack

(b)

32 mm Ø short, straight pipe with a swept connection to a 50mm diameter common or main waste

FIGURE 78

This angle must be not less than 102.5°

100 to 150mm Ø discharge or soil stack

50 mm Ø vent stack

25mm Ø vent pipe

Clearing eye

32mm diameter short straight branch waste joining a 50mm diameter main waste not more than 10m long

FIGURE 79

Soil and vent pipe may be external for 3 storeys only

Ground floor waste may discharge externally provided the end of the waste is below the grid and above the water seal. Alternatively it could connect to a side inlet on the gulley

Clearing eye

Grid

Trapped gulley

Side inlet

Warning pipe

W.C. discharge for ground floor is taken direct to the drain but the branch must be less than 6m if not vented

Hot water from refrigeration   *continued*

Anticipative problems	Considerations

	lence and some water spillage into the branch, and eventually the seal may be destroyed. See figure 81.
2 Blockage	Provide clearing eyes at the following positions and make them easily accessible. See figure 82.
	(a) On traps adjacent to the appliance if the unit does not have an integral one.
	(b) At bends and junctions on waste pipes.
	(c) On soil pipes on alternate floors at least.
	(d) Alternatively provide bottle traps where deposits may be excessive, for example, a basin in a hair dressing salon.
	(e) At the end of common wastes.
3 Health hazards	1 Provide traps for each appliance or each branch whichever is applicable. See figures 77 and 82(e).
	Note A resealing tap collects the sediment and scum which would normally be distributed amongst several individual ones, so it will require more frequent attention.
	2 Provide fresh air for the drains with a soil and/or vent pipe, terminating it above the level of openable windows. If this terminal is located within 3 m of an openable window or vent then it must extend 1 m above it. See figure 83. In situations where the S & VP is on a flank wall which has no windows the minimum height for the terminal would be about 3 m above ground level in the case of a bungalow, otherwise at least 1 m above the highest appliance.
	Note The soil and vent pipe may be external for up to 3 storeys only.
4 Noise	Locate the S & VP away from the quiet areas such as bedrooms. Disturbing noises may not only come from the discharges going down the stack but also from the refilling of the flushing cistern, especially if the water pressure is high. Where the services and quiet areas have to be close then some noises may be reduced by providing cupboards and built-in wardrobes on the separating walls.
5 Economic installations	Group appliances around the stack rather than have them spread out along the wall and possibly having to use bends in the wastes. Also arrange sanitary appliances above each other and use a common stack pipe. They need not necessarily be directly above each other provided that they can run directly into the vertical stack without the need for bends, excessive lengths and extra pipes for anti-siphonage runs. See figure 84.
	Locate the WC nearest to the S & VP wherever possible as this has the largest diameter and is the most expensive pipe.
	Note It is not possible to run a WC discharge pipe along the back of a pedestal wash basin without using a horizontal duct on the wall face or going below floor level.

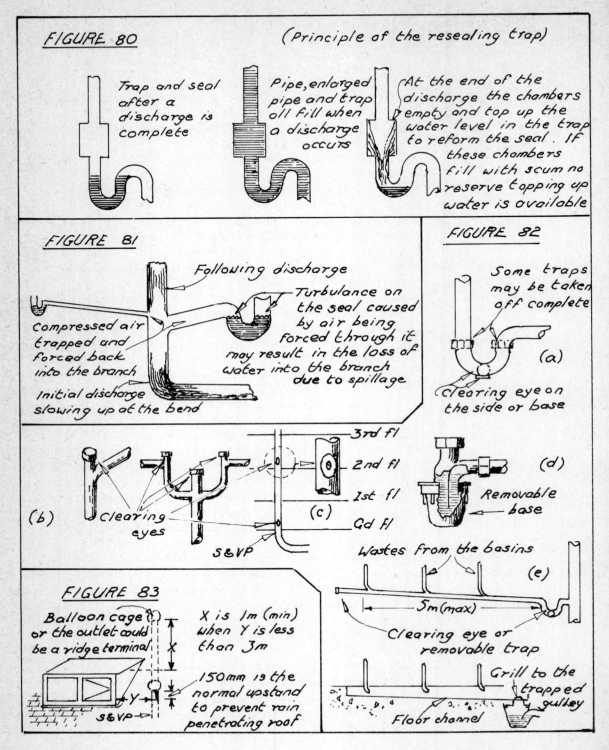

FIGURE 80 (Principle of the resealing trap)

Trap and seal after a discharge is complete

Pipe, enlarged pipe and trap all fill when a discharge occurs

At the end of the discharge the chambers empty and top up the water level in the trap to reform the seal. If these chambers fill with scum no reserve topping up water is available

FIGURE 81

Following discharge

Compressed air trapped and forced back into the branch

Turbulance on the seal caused by air being forced through it may result in the loss of water into the branch due to spillage

Initial discharge slowing up at the bend

FIGURE 82

Some traps may be taken off complete

(a)

Clearing eye on the side or base

(b) Clearing eyes

(c) 3rd fl / 2nd fl / 1st fl / Gd fl

S&VP

(d) Removable base

(e) Wastes from the basins

5m (max)

Clearing eye or removable trap

Grill to the trapped gulley

Floor channel

FIGURE 83

Balloon cage or the outlet could be a ridge terminal

X is 1m (min) when Y is less than 3m

150mm is the normal upstand to prevent rain penetrating roof

S&VP

Materials *continued*

Anticipative problems	Considerations

6 Wastes to baths and showers

Because baths and showers are low down and some wastes are very long and must fall, there is often difficulty in placing them. When appliances are over timber floors then:

(a) ensure that the wastes can run between the joists as they cannot run through them. This may require the timbers to be trimmed to form a duct as shown in figure 19(a);

(b) provide access to the trap and clearing eyes. If this is not possible from above then

(c) provide an access panel in the ceiling below ensuring that it is not over a lounge or bedroom and, in the case of flats, it is not in premises owned or occupied by other people;

(d) alternatively provide a running trap near to the stack in a common duct ensuring that the waste length does not exceed 5 m. See figure 82(e).

When waste runs are above precast concrete and cast *in situ* concrete floors then

1 if the waste is long raise the bath or shower tray on a plinth to give the necessary height for proper falls, as it cannot run through the mass concrete or precast units;

2 provide a trap and access as in **6**(b) above.

7 Bends in the main vertical stack

Avoid bends in the main vertical stack at all times. Exceptions, however, shown in figure 85 are:

(a) when the stack ceases to be a soil pipe and becomes a vent only. Any change in direction must rise at an angle not less than 12.5 degrees;

(b) below the lowest appliance connected to it, for example, where a soil pipe at ceiling level has to cross from the drop to a wall or column in order to avoid being in space as in a covered car park. When this is unavoidable then

1 ensure that the fall across the ceiling is approximately 2.5 degrees;

2 provide access covers at each end of the ceiling run on the bends, or as near as possible, to permit the clearance of probable blockages. When the space below is very high, is subject to considerable traffic or is for public use, for example a reception area, then it is better and easier for maintenance if the access plates are in the floor above. This also avoids unpleasant spillages in the area below. See figure 85;

8 Mechanical damage

3 provide protection for the pipes against mechanical damage, especially to drops in areas such as underground carparks. See figure 85.

9 Materials

Copper This is expensive, relatively light in weight, easy to prefabricate and has a long life. The connections may be capillary soldered, brazed, or, if it is anticipated that units may have to be removed, then compressed joints.

Cast iron This is a heavy but durable material and there is a very large

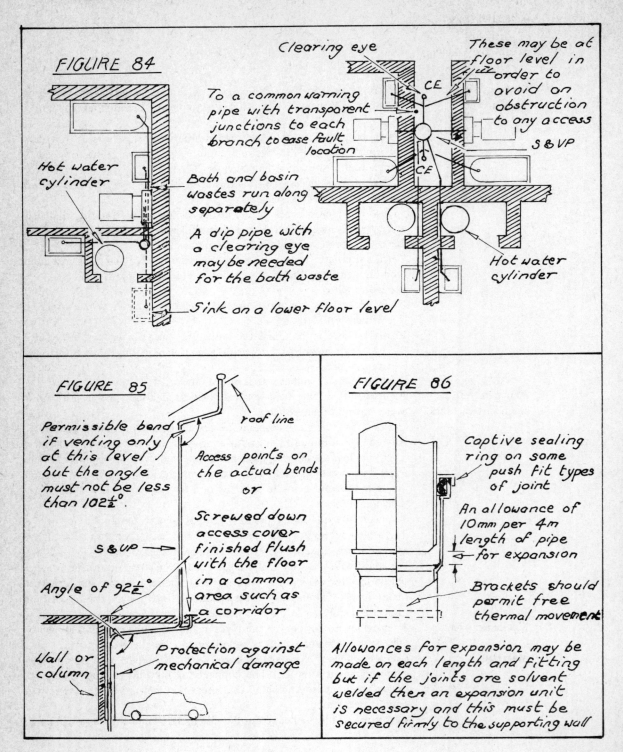

FIGURE 84

Clearing eye

To a common warning pipe with transparent junctions to each branch to ease fault location

These may be at floor level in order to avoid an obstruction to any access

CE

CE

S & VP

Hot water cylinder

Bath and basin wastes run along separately

A dip pipe with a clearing eye may be needed for the bath waste

Sink on a lower floor level

Hot water cylinder

FIGURE 85

Permissible bend if venting only at this level but the angle must not be less than $102\tfrac{1}{2}°$.

roof line

Access points on the actual bends or

Screwed down access cover finished flush with the floor in a common area such as a corridor

S & VP

Angle of $92\tfrac{1}{2}°$

Wall or column

Protection against mechanical damage

FIGURE 86

Captive sealing ring on some push fit types of joint

An allowance of 10 mm per 4 m length of pipe for expansion

Brackets should permit free thermal movement

Allowances for expansion may be made on each length and fitting but if the joints are solvent welded then an expansion unit is necessary and this must be secured firmly to the supporting wall

70

Materials *continued*

Anticipative problems	Considerations

range of fittings available. It is usually supplied coated with a bituminous solution internally and externally. Lengths are available with ears cast on for fitting direct to the wall, but this makes the job of removing a length, should it be so required, rather difficult compared with normal pipe clips.

Galvanised steel   This material is lighter than the last but if other metals are used for the wastes then there is a possibility of electrolysis affecting the zinc in the stack. Painting is recommended to give added protection from atmospheric action. Where lengths have to be cut the unprotected end should be regalvanised.

Plastics   These materials are very light, easy to handle, very resistant to corrosion, and have a large range of fittings available. Expansion however, when compared with other materials is much higher and must be provided for, especially in the vertical stack. See figure 86. Jointing is done with solvent welds, the use of 'O' rings and some have patent screwed sealing rings. The materials generally used are
1   *Unplasticised polyvinyl chloride* (uPVC)   Care is needed as the limit in temperature of the discharge must not exceed 65°C which would make it unsuitable for carrying waste from laundries.
2   *Acrylonitrile butadiene styrene* (ABS)   This material can cope with hotter liquids than uPVC.
3   *Polythene and polypropylene* – These are mostly for laboratory work. The latter can carry water of a much higher temperature.
Note   In all cases where plastics are to be used the manufacturer should be consulted regarding its use and limitations.

Flushing cisterns or water waste preventers

These appliances are used to control the amount of water run off each time the unit is operated, and so reduce wastage. The usual volume used per flush is 9 litres but the new Water Byelaws require domestic units to have a dual flush which provides either 4.5 or 9 litres depending on how the mechanism is operated. Flushing cisterns may be:
1   chain operated if fitted at high level and in the same area as the WC pan being flushed;
2   lever operated if fitted at high level and hidden in a duct, or if part of a low level suite;
3   knob operated, the actual button to be depressed being either on the cistern itself or part of a remote controller located on the wall or floor.
   The actual designs and materials vary according to the type of use for which it is selected and the kind of finish required.

Anticipative problems	Considerations

1 Choice of unit

(a) Bell or well type — These are usually fixed at high level and chain operated. See figure 87. They may be lever operated if installed in a duct behind the WC pan but access to the cistern is essential by a removable panel adjacent to it in order that maintenance work can be carried out. See figure 90.

Note This type of cistern is not permitted in new houses because they provide a single flush of 9 litres only.

(b) Diaphragm type — These may be used at high and low level, and can be obtained as a single flush unit discharging 9 litres, or a dual flush type which supply 4.5 or 9 litres as required. See figure 88. Easy access to the unit is essential for the carrying out of maintenance work or replacement.

Note Water Byelaws now stipulate that dual flush types only are to be installed in domestic premises in order to conserve water. However this would not be suitable for other establishments at present because, although instructions on how to use them may be displayed, it is doubtful if the public would stop to read them.

(c) Trough type — In some public areas, for example a busy railway station, consecutive flushes of the toilet may be required in less time than it takes to refill the cistern, in which case either the pan is left fouled or the user suffers unnecessary frustration waiting for the filling to complete. The use of trough cisterns with their large reservoir overcomes this problem. See figure 89. *Note*:

1 The trough and operating rods may be hidden in ducts behind the WC thus exposing only the pans and operating level. See figure 90.

2 Only one warning pipe is required.

3 Only one ballvalve is required but should it fail then all the compartments served by the unit will be out of use. It is better to have two of these troughs for each sex in any toilet area used by the public, employees, etc, in order that one set is always available for their use.

4 Access to the trough and flushing mechanisms is essential as stated before.

(d) Automatic flushing cisterns — These are essential for gentlemen's urinals but unless controlled will operate for 24 hours each day irrespective of whether the building is in use or not. The principles involved are shown in figure 91 but the Water Byelaws now require that some form of mechanism be provided to prevent this obvious waste of water. This might be a time switch which puts the unit into operation only when the building is occupied, or it could be a hydraulic valve which is activated by the pressure variations resulting from people in the building using any of the sanitary appliances installed. See figure 92. When the building is unoccupied then the forces on either side of the valve will remain constant and so it will stay closed. If any tap or flushing cistern is later used the water flowing from the outlet will cause a small reduction of pressure in the pipes on

72

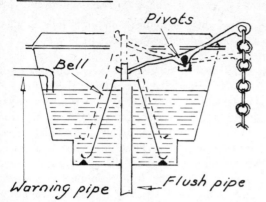

FIGURE 8T

Bell or Well type cistern

Possible faults may be

(1) Failure of the ballvalve (not shown)

   (a) washers need replacement

   (b) seating may become grooved

   (c) float may be punctured.

(2) Pivots wear and the bell is not then raised high enough to start the flushing action

Labels: Pivots, Bell, Warning pipe, Flush pipe

FIGURE 88

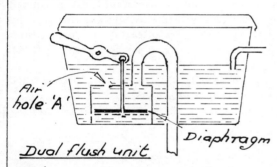

Piston or diaphragm type cistern

Possible faults may be

(1) Ballvalves as stated above

(2) Diaphragm may split or perish and need to be replaced

(3) Any small perforations in the 'U' bend will prevent the siphonic action working

Labels: Air hole 'A', Dual flush unit, Diaphragm

Note:– If the lever is depressed and held down sealing off the air inlet at 'A' then 9 litres of water will be flushed from the unit. If the lever is operated and released as soon as the flush starts then air entering the chamber at A will stop the siphonic action once 4.5 litres have left the flushing cistern

FIGURE 89

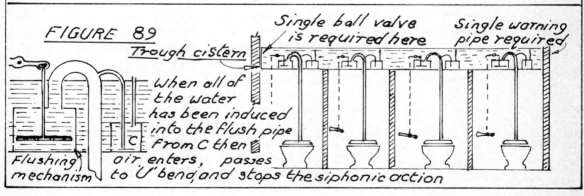

Trough cistern

Single ball valve is required here

Single warning pipe required

When all of the water has been induced into the flush pipe from C then air enters, passes to 'U' bend, and stops the siphonic action

Labels: Flushing mechanism, C

73

Flushing cisterns or water waste preventers *continued*

Anticipative problems	Considerations
	one side of the valve, destroying the equilibrium which existed before, resulting in the valve opening and allowing water to flow to the automatic flushing cistern.
2 **Warning pipes**	Provide warning pipes to all flushing cisterns except automatic ones and arrange for the discharge points. See figures 32 to 37 inclusive.
	Note The discharge of warning pipes into gulleys is not recommended with underfloor heating, as the warmth generated during the winter will cause the water seal to evaporate and permit the smells and gases from the drains to permeate into the building. If there is no alternative then the problem may be reduced or even eliminated by using deep sealed gulleys and connecting a waste from a frequently used wash basin to a side inlet on it. This will constantly top up the seal but provisions for clearing the waste are needed.
3 **Access to the unit**	Provide access to the cistern and ensure that it is large enough to allow easy maintenance and the replacement of the unit if and when required.
4 **Pressure**	1 Provide a minimum head of 1m for normal flushing cisterns if the feed pipe is not already serving other appliances. For automatic cisterns with hydraulic controls then seek the manufacturer's advice as it may be more. See figure 92.
	2 Ensure that the correct ballvalve for the pressure anticipated is fitted. (See automatic water supply to cisterns in section 1.)

Sanitary appliances

These vary considerably in size, type, use, finish, etc, and the following comments concern some of the basic units together with variations which are available.

Anticipative problems	Considerations
	Units for the removal of human excreta
1 **Choice of WC** (a) High level suite	The cleansing relies entirely on the wash down action of the flow. Generally the whole unit is exposed so it is not recommended for some areas in mental institutions, or where anti-vandal installations are required unless the cistern and the chain can be screened. See figure 93(a).
(b) Low level suite	These may be cleansed by wash down or by combination of the flush and siphonic action. See figures 93(b) and (e). The cistern may be exposed, or if space is available, hidden in a duct.
(c) Close couple suites	The cistern sits on the back of the pan and the unit operates by siphonic action started by the flush. See figures 92(c) and (f).
	Note Siphonic units are more expensive than other types but are neat, generally quieter and more hygienic. The authorities may prohibit these being connected direct to the mains.

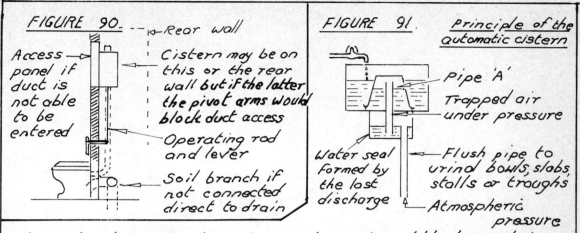

FIGURE 90.

Access panel if duct is not able to be entered

Rear wall

Cistern may be on this or the rear wall but if the latter the pivot arms would block duct access

Operating rod and lever

Soil branch if not connected direct to drain

FIGURE 91. Principle of the automatic cistern

Pipe 'A'

Trapped air under pressure

Water seal formed by the lost discharge

Flush pipe to urinal bowls, slabs, stalls or troughs

Atmospheric pressure

(1) Water drips into the cistern continuously and the trapped air between the freshly added water and the seal formed by the previous discharge becomes compressed

(2) The pressure increases as the fill continues and when it becomes more than atmospheric some compressed air is forced through the water seal. As the air is released so the pressure in pipe 'A' is reduced to a level less than atmospheric. Normal air pressure on the surface of the water now being greater than that in the pipe forces water into pipe 'A' and siphonic action empties the cistern.

(3) The control must provide a flush of 4.5 litres to each bowl or 700mm run of slab at intervals of not less than 20 minutes.

Note:- Ball valves and warning pipes are not required for automatic flushing cisterns.

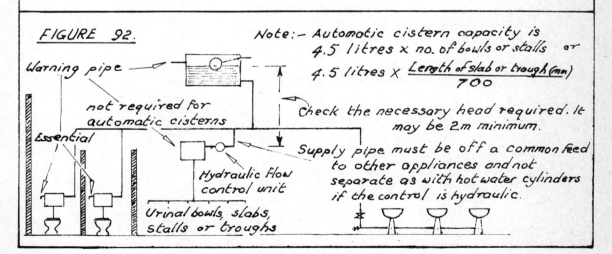

FIGURE 92.

Warning pipe

not required for automatic cisterns

Essential

Hydraulic flow control unit

Urinal bowls, slabs, stalls or troughs

Note:- Automatic cistern capacity is 4.5 litres × no. of bowls or stalls or
$$4.5 \text{ litres} \times \frac{\text{Length of slab or trough (mm)}}{700}$$

Check the necessary head required. It may be 2m minimum.

Supply pipe must be off a common feed to other appliances and not separate as with hot water cylinders if the control is hydraulic.

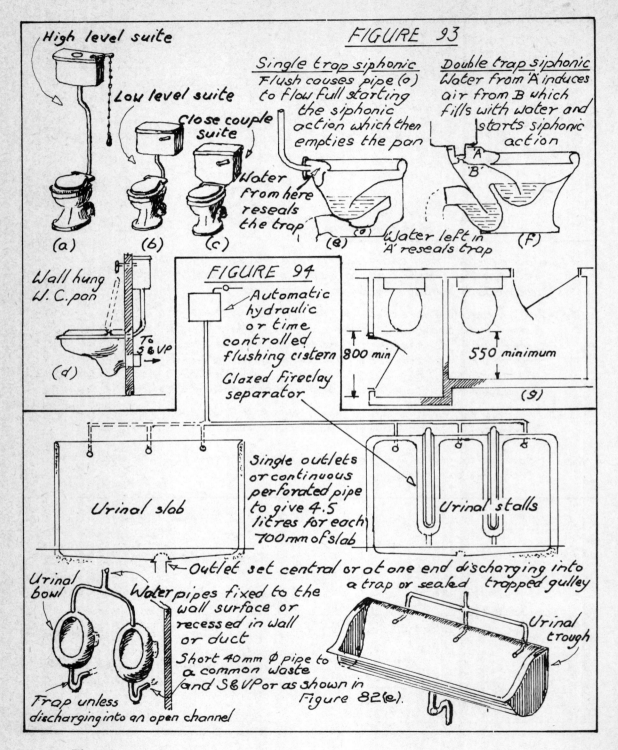

High level suite

Low level suite

Close couple suite

(a)   (b)   (c)

FIGURE 93

Single trap siphonic
Flush causes pipe (o) to flow full starting the siphonic action which then empties the pan

Water from here reseals the trap

(e)

Water left in 'A' reseals trap

Double trap siphonic
Water from 'A' induces air from B which fills with water and starts siphonic action

A
B

(F)

Wall hung W.C. pan

To S&VP

(d)

FIGURE 94

Automatic hydraulic or time controlled flushing cistern

Glazed Fireclay separator

800 min

550 minimum

(9)

Single outlets or continuous perforated pipe to give 4.5 litres for each 700mm of slab

Urinal slab

Urinal stalls

Outlet set central or at one end discharging into a trap or sealed trapped gulley

Urinal bowl

Water pipes fixed to the wall surface or recessed in wall or duct

Short 40mm Ø pipe to a common waste and S&VP or as shown in Figure 82(e).

Urinal trough

Trap unless discharging into an open channel

Water Closets *continued*

Anticipative problems	Considerations
(d) Wall units	Some pans are designed to be fixed directly on to the wall, on brackets, leaving the floor area below absolutely clear and so avoiding the difficult job of cleaning around the outlet pipe behind the pan. See figure 93(d).
(e) WCs with timber insets	These have timber insets on the rim and are good where hard, rough usage is anticipated but are not as hygienic as those fitted with drop down seats and covers.
(f) WCs with integral clearing eyes	In situations where blockages may be anticipated to be a problem, for example in schools due to an excessive use of paper, then pans with an integral clearing eye are advisable. This might be advantageous also in places where it is difficult to provide access at the end of a common soil pipe collecting discharges from several WCs.
2 Space around the unit	Provide a minimum space of 550 mm between the wall and the front of the pan, or 800 mm if fitted behind a door in order that on swinging open it does not damage the pan or seat. See figure 93(g).
3 Underground or underfloor pipes	Arrange the WCs so that the pipes from them go direct to the S & VP or, in the case of ground floors, to the drain. Avoid layouts which require soil pipes or drains to cross each other, because as they start and finish at similar levels and must run straight it is difficult and most times virtually impossible. If not considered early on then alterations will be required at a later date which could lead to further problems which might never have arisen. See combined drains Figure 136.
4 Choice of urinal fittings	These fall into one of four categories as illustrated in figure 94 and include:
	(a) Bowls These are wall mounted and corner models are available. Concealment of the pipes provides a smarter finish, and generally it is more hygienic in that less area is exposed to any possible fouling and the surfaces around are more easily cleaned.
	(b) Troughs These are similar to the last in that the floor is unobstructed except for a single waste which may be exposed, but are more economic as the amount of waste pipe required is much less.
	(c) Stalls These offer some small measure of privacy but are less hygienic than slabs because of the much larger surface area to be kept clean, and the junctions of the separators and the slab where germs can collect in the joint.
	(d) Slabs As the name suggests this offers one large flat surface with few or no joints to harbour germs. As everything is carried away in floor channels there is no problem of exposed wastes.
5 Discharge from the urinals	1 Provide traps at the end of all floor gulleys before connecting them either to the S & VP or, in the case of ground floor appliances, the drain.
	2 Provide a common waste for a range of urinal bowls but, if possible, in a duct to avoid runs below the units which may be fouled and diffi-

Urinals *continued*

Anticipative problems	Considerations
	cult to clean. When units do not have individual traps then provide a running trap just prior to the S & VP and ensure that it is easily accessible.
6 **Leaks from the appliances**	Provide an impervious membrane under slabs and stalls on upper floors as the jointing of them and the channels cannot be guaranteed watertight.
7 **Anticipated users**	Ensure that the appliances installed can be used by all the people for whom they were intended. *Note* If a young boy out shopping with his parents wishes to use the public car park toilet he may find that urinal bowls or troughs have been installed. Being small he is not able to reach them, whereas if slabs or stalls had been provided there would have been no problem.
8 **Slop sinks and bed pan washers**	Provide hot and cold water taps and a high level flushing cistern. Some units are fitted with sprays to assist the cleaning. See figure 95.
9 **Isolation and drainage of the branches**	Provide isolating valves for ranges of WCs and urinals but ensure that all are not out of use at any one time. This applies particularly to the trough type cistern which serve several cubicles but is dependant upon a single ballvalve. Drain valves are not required for pipes which drop and terminate at the ballvalve as they may be emptied by flushing the unit. If they extend to the floor and then rise to the ballvalve then drainvalves must be provided at the lowest point.

Sanitary appliances for personal ablutions

1 **Choice of wash basin**	This again is dependant on the particular design, colour, material and finish required. It may be a single unit supported on:
	(a) exposed wall brackets some of which have an additional rail for hand towels. See figure 100;
	(b) concealed wall brackets. This will require a separate provision for hand towels which might be a rail linked in with the heating system to supply a little background warmth in the room;
	(c) a matching pedestal which looks neat in that it screens the vertical pipes. Concealed horizontal runs may be in a duct behind the pedestal or run below floor level but this means that:
	    1 in timber floor construction the pipes must run between the joists. See figure 101;
	    2 with solid floors the screed must be thick enough to conceal them which may be much thicker than anticipated if they have to cross.
	(d) It may be recessed into a flat chipboard or timber slab finished with a smart impervious plastic laminate. The joint is sealed with a cover strip but as the surface is level there is no provision for spilt water to drain back into the basin. See figure 102;

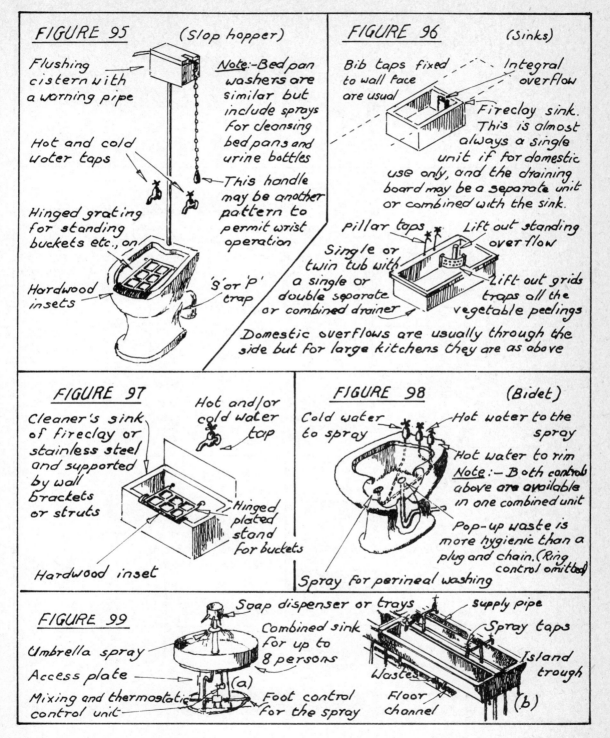

FIGURE 95 (Slop hopper)

Flushing cistern with a warning pipe

Hot and cold water taps

Hinged grating for standing buckets etc., on

Hardwood insets

Note:-Bed pan washers are similar but include sprays for cleansing bed pans and urine bottles

This handle may be another pattern to permit wrist operation

'S' or 'P' trap

FIGURE 96 (Sinks)

Bib taps fixed to wall face are usual

Integral overflow

Fireclay sink. This is almost always a single unit if for domestic use only, and the draining board may be a separate unit or combined with the sink.

Pillar taps

Lift out standing overflow

Single or twin tub with a single or double separate or combined drainer

Lift out grids traps all the vegetable peelings

Domestic overflows are usually through the side but for large kitchens they are as above

FIGURE 97

Cleaner's sink of fireclay or stainless steel and supported by wall brackets or struts

Hot and/or cold water tap

Hinged plated stand for buckets

Hardwood inset

FIGURE 98 (Bidet)

Cold water to spray

Hot water to the spray

Hot water to rim

Note:- Both controls above are available in one combined unit

Pop-up waste is more hygienic than a plug and chain. (Ring control omitted)

Spray for perineal washing

FIGURE 99

Umbrella spray

Access plate

Mixing and thermostatic control unit

Soap dispenser or trays

Combined sink for up to 8 persons

Foot control for the spray

(a)

supply pipe

Spray taps

Island trough

Wastes

Floor channel

(b)

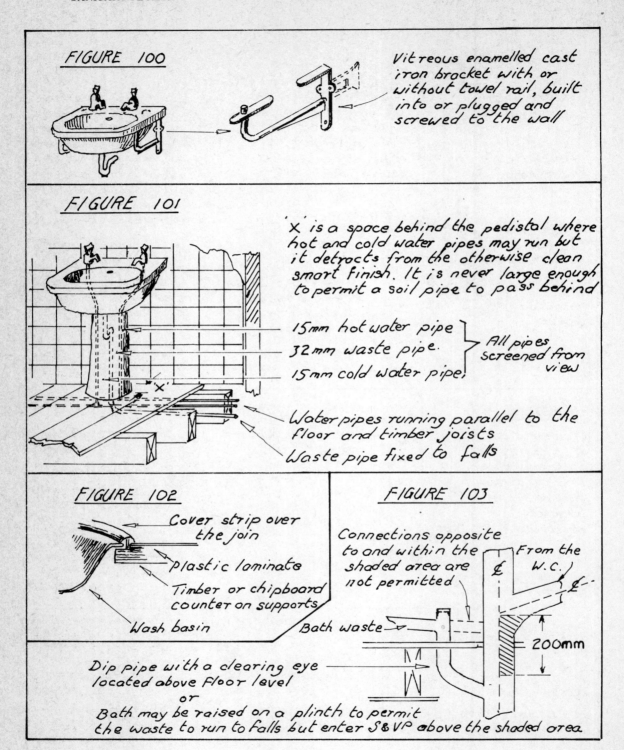

FIGURE 100

Vitreous enamelled cast iron bracket with or without towel rail, built into or plugged and screwed to the wall

FIGURE 101

'X' is a space behind the pedistal where hot and cold water pipes may run but it detracts from the otherwise clean smart finish. It is never large enough to permit a soil pipe to pass behind

15mm hot water pipe
32mm waste pipe.
15mm cold water pipe!

All pipes screened from view

Water pipes running parallel to the floor and timber joists
Waste pipe fixed to falls

FIGURE 102

Cover strip over the join
Plastic laminate
Timber or chipboard counter on supports
Wash basin

FIGURE 103

Connections opposite to and within the shaded area are not permitted
From the W.C.
200mm
Bath waste

Dip pipe with a clearing eye located above floor level
or
Bath may be raised on a plinth to permit the waste to run to falls but enter S&VP above the shaded area.

Sanitary appliances *continued*

Anticipative problems	Considerations
	(e) Sometimes the units are completely boxed in to hide the pipes when ducts are not possible or practical. In this case access panels must be provided in order to get to the traps, clearing eyes and tap connections in case at any time they need to be cleared or replaced.
	Note Considerations as mentioned under single units also apply when a range of fittings is installed. However difficulties arise in that larger diameter common wastes are required and in many instances some venting measures to avoid siphonage.
	(f) Where provision for many persons is required but appearance is of lesser importance, for example in schools and factories, then communal units may be specified such as ablution fountains. See figure 99(a), and island trough sinks. See figure 99(b).
2 Space around the unit	Provide a minimum of 650 mm between the basin and any wall facing it. When a range of units is installed provide a minimum space of 1 m in order to allow room for people to pass.
3 Choice of bath	This will depend generally upon the colour, shape and finish required with consideration also to its location and weight. See materials.
4 Location of the bath	1 Ensure that very heavy fireclay baths are at right-angles to the run of the joists in order that the maximum number of timbers are providing support. The weight when in use may be several hundredweights.
	Note 1 cwt is approx. 50 kg
	2 Check to see if the fall of the bath waste will clear the WC outlet at the S & VP or whether a dip pipe and clearing eye is required. See figure 103.
	3 If the run above the floor is not possible or a dip pipe is not approved then either
	(a) raise the bath on a plinth to give the extra height but provide for some adjustment to the finishes for the exposed step or
	(b) drop the waste below the floor and let it run between the joists. Sometimes the S & VP can be moved to occupy the same void as the bath outlet between the joists, but care is needed to ensure that problems with other pipes are avoided.
5 Fixing of bath panels	Ensure that any bath panels, once fixed, can be easily removed. If walls are to be tiled then ensure that the tiles run behind the edge of the bath panels and not the panels behind the tiles. If a pipe leaks and tiles have to be stripped off the wall before the panels can be removed then a small plumbing repair can finish up as an expensive, major operation. See figure 104.
6 Choice of shower (a) *In situ* showers	*In situ* showers are usually for a number of units and constructed on site. They consist of a concrete slab laid to fall towards the floor channel which discharges into a trapped gulley, the walls and floor being finished with glazed tiles or other easily cleaned, hard wearing,

FIGURE 104

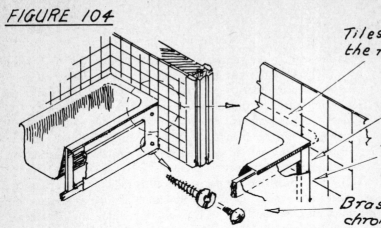

Tiles brought down over the rim of the bath

Timber framing fixed in front of the tiles or given a recess for them

Bath panel screwed to the timber frame when tiling is complete

Brass screws with removable chromium plated dome caps

FIGURE 105

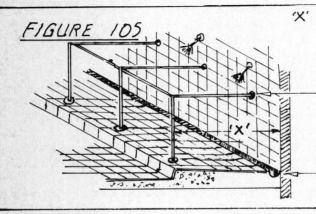

'X' is 900mm minimum for cubicle showers but 1200 when it is a run through shower

Partitions 1.8m high starting 150mm off floor level or rails carrying curtains which may pull along the front or side as needed

Channel with a grid over and falling to a trapped gulley

FIGURE 106

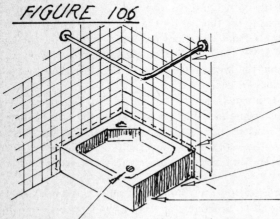

Curtains run around on two sides or could be replaced with glazed screens on one or both sides

Tiles brought down over the edge of the unit to prevent water penetration between the face of the tiles and the side of the tray

Lead tray inserted below the shower tray and turned up wall

600 up to 900 mm square ceramic or plastic shower tray

Outlet to a floor channel or direct to a waste pipe. Note:- Once the tray is fitted then the pipes below are inaccessible

Sanitary appliances *continued*

Anticipative problems	Considerations
	impervious material. Often they are separated into cubicles with rails and curtains, or solid partitions finished to match the floors and walls. See figure 105.
(b) Built in showers	Complete ceramic or plastic shower trays are purchased and installed in prepared recesses, the wall finishes being applied *in situ* and brought down over the edge of the tray to cloak the joint between the tray and wall. See figure 106.
(c) Packaged shower unit	These are factory made units including the tray, screen walls and possibly the service pipes and fittings. Sometimes units are combined, with the bath acting as the shower tray.
7 Water pressure	Ensure that the hot and cold water taps have the same head otherwise mixing will not occur due to the pressure variations. Ensure that the minimum head above the highest position of the spray is not less than 1 m. Where the supply head is less than 1 m then the flow may be pump assisted. See figure 15.
8 Height of the shower head	People are of different heights and have varying habits, for example some may wish to stand under the shower getting completely wet whilst others prefer to keep their heads dry. Where fixed outlets are not required provide handsprays which can be detached or attached to a sliding fitting on a fixed bar if so required. See figure 107.
9 Space in shower	This is usually fixed by the size of the tray but for larger units allow a minimum of 900 x 900 mm unless it is a run through unit which will be a depth of 1200 mm. See figure 105.
10 Leaks, especially on upper floors	Joints below the tray cannot be guaranteed to be watertight so provide an impervious membrane below the unit to prevent any seepage downwards.
11 Blockages of the waste	Provide a running trap if access to the underside of the unit is not possible. See figure 108.
12 Bidets	These are becoming more popular in this country. If required then hot and cold water supplies must be provided together with a waste. See figure 98.

Units for the preparation of food

1 Choice of sink	Variations in the type of sink usually depend on whether they are for general use or for special purposes, as required in hospitals and restaurants.
(a) Domestic sinks	These may have single or twin bowls and single or double draining boards. Units which have combined draining boards are much more hygienic than the ones with detached draining boards. The overflow in fireclay sinks is usually moulded in the sink itself and more easily cleaned than the one in metal sinks which is a hole in the side behind a grill leading to an attached flexible tube connected direct to the sink waste. See figure 96.

Sanitary appliances *continued*

Anticipative problems	Considerations
(b) Vegetable sinks	These are larger and usually for commercial use. As the amount of food preparation is greatly in excess of domestic use grids are provided to prevent peelings, etc, getting into the waste and blocking the trap. Also an integral overflow would be difficult to keep clear and clean so standing overflows in the form of a straight pipe which can be lifted out, are provided. See figure 96.
2 Space around unit	Provide a minimum space of 450 mm standing room in front of the sink but if passing room is needed, as in large kitchens, then this should be extended in length to at least 1100 mm.

Units for the washing of utensils, crockery, mops and cloths

1 Appliances	These may be as already mentioned under sinks, the unit having to provide a dual function. In addition some of the following may be required for larger establishments and kitchens.
(a) Cleaner's sink	This is important for offices, schools and the larger buildings. It should be in a special area reserved for cleaners and their equipment rather than expecting them to wash their cloths and mops and clean their buckets in other units in the general toilet or kitchen areas. See figure 97.
(b) Rinsing sink	These are used in large kitchens and incorporate a built in heater often requiring a three phase electricity supply. The water is too hot for hand use and the cups, plates, etc, are placed in trays and dipped before drying.
(c) Sinks for glassware	These may be of timber or have timber inserts or trays in the bottom. Care is still essential when washing tumblers but the timber reduces the noise and is more resilient than other materials and less likely to cause breakages.
2 Location of the sink	Generally there is no problem as it is usually located near to the S & VP serving other appliances, but when odd units, such as cleaner's sinks and those serving small bars, have to be considered they can end up as a very costly installation. When deciding where to put these units: 1 try to locate them near to although separate from other areas using sanitary appliances, and in the case of upper floors never more than 5 m from a S & VP. In the case of ground floor units the waste could be taken direct to a drain but the authorities would ask for a rodding eye near the appliance to permit the drain to be cleared if blocked. See figure 109; 2 check that the location is such that the run from the sink to the main drain does not exceed 6 m otherwise a separate vent pipe will be needed at the top end of that branch drain or, in the case of upper floors, an additional S & VP and in both cases it will have to be taken to high level. See figure 110.

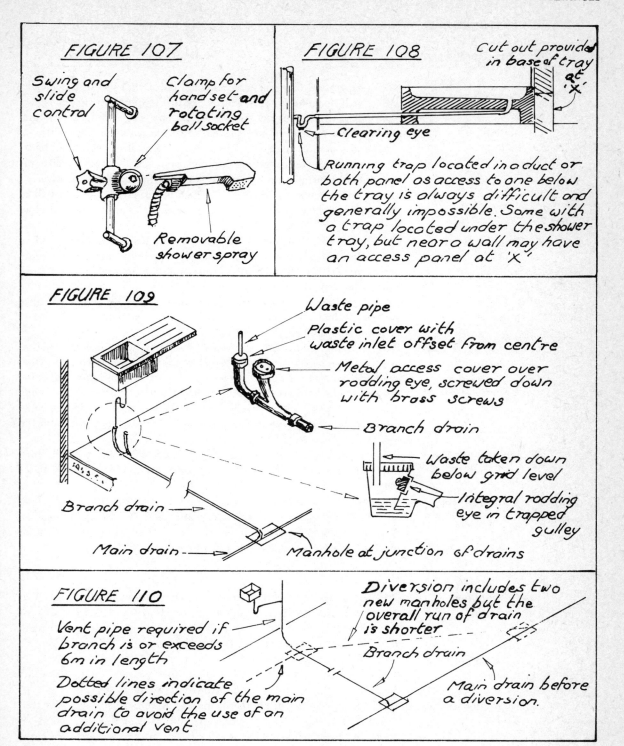

FIGURE 107

Swing and slide control

Clamp for hand set and rotating ball socket

Removable shower spray

FIGURE 108

Cut out provided in base of tray at 'X'

Clearing eye

Running trap located in a duct or both panel as access to one below the tray is always difficult and generally impossible. Some with a trap located under the shower tray, but near a wall may have an access panel at 'X'

FIGURE 109

Waste pipe

Plastic cover with waste inlet offset from centre

Metal access cover over rodding eye, screwed down with brass screws

Branch drain

Waste taken down below grid level

Integral rodding eye in trapped gulley

Branch drain

Main drain

Manhole at junction of drains

FIGURE 110

Diversion includes two new manholes but the overall run of drain is shorter

Vent pipe required if branch is or exceeds 6m in length

Dotted lines indicate possible direction of the main drain to avoid the use of an additional vent

Branch drain

Main drain before a diversion.

Sanitary appliances *continued*

Anticipative problems	Considerations
	3 If a separate S & VP is accepted then check that the drop coincides with a wall or column below and is not in the middle of a used space. *Note* When a single unit is located a long way from other units, although the item itself may be cheap the expense for the additional work and materials such as drains plus the excavation, concrete, back-fill, manholes and extra venting required can make the cost very difficult to justify the installation. In some cases the main drain may be diverted nearer to the appliance concerned which is still costly but can reduce the length of the branch and so avoid the need for and extra vent pipe. See figure 110.
3 **Draining boards** (a) Glare	Glare from white enamelled or mirror finished stainless steel draining boards located under windows exposed to the sun may bother some users whereas tinted and satin finishes may be no problem and more acceptable in this situation.
(b) Breakages and noise	Hardwood timber draining boards are quieter and less prone to cause breakages than other materials, but they are less hygienic and need to be treated periodically to extend their life.
(c) Shrinkage and movement	Prior to fixing, timber draining boards should be kept for a short period in the same environment as they will be fixed, in order that any antici-pation shrinkage can occur. Large wall tiles have been known to come off due to the hardwood draining board in contact with them shrinking after occupation. Alternatively a non-adhesive plastic or metal sepa-rating strip placed between the timber and the tiles may provide for the independent movement necessary.
4 **Washing machines**	Partly plumbed in portable units require a waste upstand 600 mm long, minimum, above the trap and left open to receive the machine's outlet hose.
5 **Tumble dryers**	These should be located on outside walls in order that the warm, moist air can be taken direct to the external atmosphere. The use of portable, flexible hoses to open windows and doors from units sited in other locations is an inconvenience best avoided.
	Special purpose appliances
(a) Drinking fountains	Drinking water is essential for all buildings but in many, other than domestic, drinking fountains may be the most satisfactory way of providing it. Each fountain must receive water direct from the main and each unit will require a waste pipe and trap.
(b) Hand rinsing basins	These are smaller than normal wash basins and fitted near WCs. In public areas the units are often built into the wall and the on/off operation of the water supply controlled by the feet.
(c) Hand basins for invalids	These are similar to normal basins but may have pop up wastes operated with the feet, and taps which are turned on by using the wrist, elbow or knee. In hospitals standing overflows are used for surgeon's sinks

Sanitary appliances *continued*

Anticipative problems	Considerations
	because they are more hygienic and the basins themselves are fitted 50 mm clear of the wall to avoid joints which are difficult to keep clean and may harbour germs.
(d) Soap dispensers	These may be operated by tilting or pressing a plunger but some are gravity fed to special taps from a storage tank located at high level. These have advantages over soap tablets in that: 1 it is more economic. It cannot be wasted by leaving it in the water, or lost due to it being taken, possible by the previous user; 2 the amount used is controlled. The ejector or dispenser will only pass a limited amount per use; 3 it is more hygienic. Each operation provides a fresh supply uncontaminated by previous users.
(e) Hand dryers	These are in various forms and may be: 1 paper towels in rolls or already cut and stacked in containers. A wall or floor mounted container is also necessary for the depositing of used papers before actual disposal; 2 continuous roller cloth towels. Whilst these are designed so that each person uses a fresh clean section, in practice the rollers can jam, the towels tear and the unit becomes temporarily unusable. Cross contamination is always possible and this danger, if present, is increased during breakdowns when several people tend to use the length of towel left exposed; 3 electrically powered warm air blowers. These are hygienic and avoid the need for towel stores and dispensers for used papers but the drying action never feels very positive and is rarely achieved in one operation of the motor.
(f) Kitchen equipment	1 Check whether commercial units, for example some steamers, require a separate mains feed to their own small feed storage cisterns located above but in the same room. 2 Check whether units such as boiling pans and bain-maries etc need to be drained of hot water where provisions of trapped gullies in tiled floor trays might be advantageous, if only to prevent possible serious accidents resulting from people having to carry buckets of hot water for disposal across the room to a normal sink.
Care on site	Provide protective covers on sanitary equipment when fixed or ensure that the rooms in which it is installed is kept locked if not in use by the workmen carrying out their normal duties. *Note* Too often sanitary appliances on site are permitted to be used by anyone on the site but no one has the responsibility for keeping them clean. This is unnecessary if the proper facilities are provided, and the stains, paint, cement, etc, so often seen on appliances during the work in process stage should never have been there, and does create a bad image to the prospective clients.

Materials for sanitary appliances

Fireclay   A hardwearing but semi-porous material so needs to be glazed. The glaze will chip and crazes with age due to the different rates of expansion between the glaze and the fireclay backing. It is heavy.

Vitreous china   A non-porous material but more expensive than fireclay.

Note: Both of the above are obtainable in white or colours, are easily cleaned, and will not distort but can break due to impact.

Cast iron   A heavy material which will not distort, and finished on exposed surfaces with porcelain enamel. It is cheaper than ceramic ware but unprotected metal in contact with water, for example the inside of flushing cisterns, will rust and this can affect its working.

Stainless steel   A hard wearing material which is smart in appearance, will not chip, is easily cleaned, and is useful for matching other non-sanitary appliances when used with kitchen cooking equipment.

Pressed steel   A hard wearing material finished with a vitreous enamel which can scratch and chip. Once the metal is exposed to water then rust formation will develop much faster than with cast iron. It can dent under impact but will not break.

Plastics   This material is very light in weight so easily handled. The colour goes right through but the surface is easily scratched and will burn but these faults can be rectified within limits. Units warm up quicker than with other materials but large items such as baths may distort a little so need to be provided with a timber frame for support which can also act as grounds for fixing bath panels.

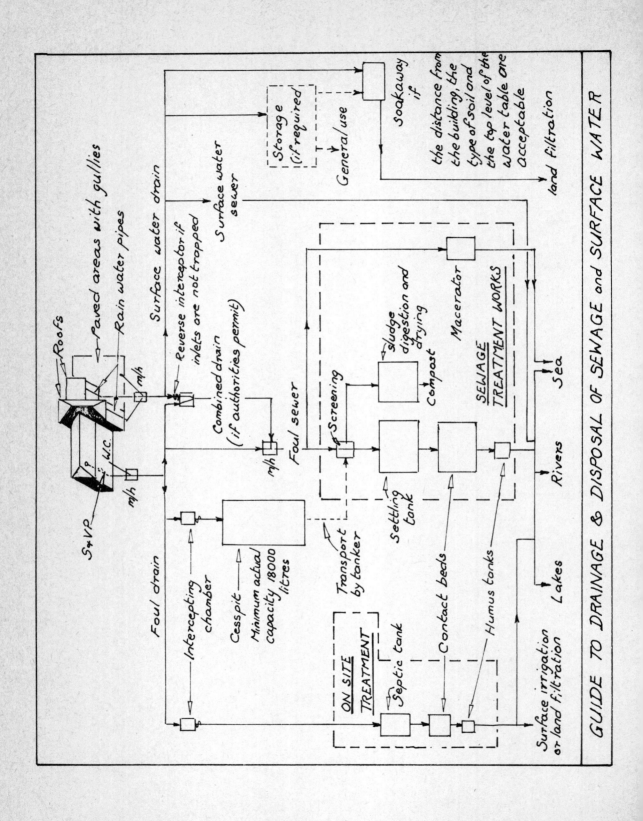

GUIDE TO DRAINAGE & DISPOSAL OF SEWAGE and SURFACE WATER

4 Foul drainage

Drainage describes a system of pipes used to collect sewage, waste water and surface water from buildings and convey it to a point for disposal. It may be put into one of three categories, these being:

(a) foul drains which carry sewage and waste water to 1 cesspits for subsequent removal by tankers which then take it to a public sewage works for treatment and final disposal; 2 septic tanks for on site treatment before disposal on or near the location of the plant; 3 foul sewers for treatment at a sewage works before passing it back into rivers, streams, lakes, etc;

(b) Surface water drains, as described in Section 6, which carry rain water from roofs and paved areas to 1 soakaways if the site and soil conditions permit it; 2 surface water sewers if they are available, or streams.

(c) Combined drains which carry both sewage and surface water to sewage treatment plants provided that both can be coped with satisfactorily and the authorities permit it.

Trenches for drains

In some excavations water may seep into the trench whilst work is proceeding and this can slow down the operation considerably and may even stop the work completely. Where problems resulting from water in soil can be reasonably anticipated then provision for directing it to sumps and pumping it out must be allowed for in the estimates.

Another item which must be considered is the provision and fixing of timbers to support the trench walls whilst work is being carried out. This will vary from struts at 2 m intervals across the trench in firm ground and runners in sandy soil to close boarded walings for those trenches excavated in clay. Also land drains are required in ground where the water table is high. See figure 113.

Where rigid pipes are used then the width of the trench can have an adverse effect on the drain in that if it is too wide then it will increase the loading on the pipes. BRS Digest 130 table 3 shows permissable trench widths for particular types and diameters of pipes together with the minimum and maximum recommended depth of cover. It also states the transition depth for each trench, this being the distance measured down from the surface to a level below which the trench must not be wider than the figure stated. If it is wider than the extra loading which might result could cause the failure of the pipe. See figure 111.

Anticipative problems	Considerations
1 **Excavation generally**	1 Ensure that the bottom of the trench follows a regular and proper fall so that any materials used for bedding will be of a uniform thickness, and the subsequent flows will achieve a self-cleansing velocity. 2 Ensure that any spots in the trench bottom which have been excavated too deep by mistake are backfilled with the same material, and well consolidated to provide a bearing surface of the same load capacity as the undisturbed soil.

Trenches for drains *continued*

Anticipative problems	Considerations

3 Ensure that the widths of any trench dug out below the level of the transition depth do not exceed the dimensions stipulated for trench widths for rigid pipes. See figure 111.

4 Where pipes have to be placed direct on to the soil ensure that

(a)   the trench bottom is dry;

(b)   there are no large stones near the surface which could form hard spots and variable bearing capacities.

Bedding for the pipes

Bedding can be divided into three classes

1   Class D bedding used for light loads where the pipes are placed directly on the natural trench bottom. The site must provide relatively dry working conditions, and the soil in the trench bottom must be reasonably soft, fine grained and uniform in consistency. It must also be capable of being trimmed by hand. See figure 112.

2   Class B bedding which is stronger and better than 1 in that the pipes are placed on a bed of granular material which in turn is brought halfway up the sides of the pipe. This practice is in general use and is advantageous on wet sites and where speed of construction is an important factor. See figure 113.

3   Class A bedding is used where strength is required and provides a concrete cradle which carries the pipe, and concrete haunching which is taken up each side of it. This type of bedding would be used for rigid pipes when placed under buildings and roads and is also useful for

(a) where the rate of slope needs to be set more accurately and maintained.

(b) trenches where it is not practical to remove any support timbering to the trenches until after the bedding has been completed.

(c) preventing any disturbances of the drains which may occur resulting from alterations at a later date requiring trenches adjacent to the existing pipe runs, and especially should they need to be dug out to a lower level.

Class A bedding is illustrated in figure 114.

There are variations in these types of bedding in that if Class D was required but the site was wet and the job did not justify the use of Class B bedding then the trench bottom could be blinded with 75 mm of granular fill. This would prevent the bottoming being destroyed by the formation of mud caused by the workmen's boots during operations.

Class B bedding may also be strengthened by surrounding the pipe in Type A material if the work did not justify Class A bedding, the filling being taken above the pipe for a depth not less than that on which it rests. Figures for trench widths, transition depths and cover depths for a variety of rigid pipes in different situations are given for Class A, B and D beddings in BRS Digest 130, table 3. An extended table is shown in the Transport and Road Research Report 303 including the variations referred to above and called Class F and S respectively. There are however differences in the figures shown on the two papers. A classification of terrain is given in BRS Digest 130 table 2. When water seepage into the trench during operations becomes a problem then a temporary drain must be provided to lower the water table at that point and prevent a flow through the actual bedding.

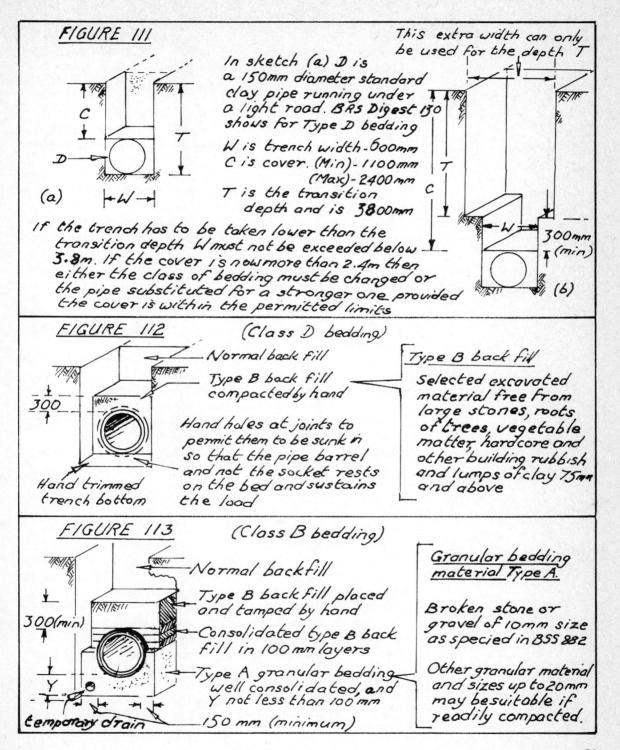

FIGURE 111

In sketch (a) D is a 150mm diameter standard clay pipe running under a light road. BRS Digest 130 shows for Type D bedding

W is trench width-600mm
C is cover. (Min)-1100mm
             (Max)-2400mm
T is the transition depth and is 3800mm

This extra width can only be used for the depth T

(a)

(b) 300mm (min)

If the trench has to be taken lower than the transition depth W must not be exceeded below 3.8m. If the cover is now more than 2.4m then either the class of bedding must be changed or the pipe substituted for a stronger one provided the cover is within the permitted limits

FIGURE 112   (Class D bedding)

Normal back fill

Type B back fill compacted by hand

Hand holes at joints to permit them to be sunk in so that the pipe barrel and not the socket rests on the bed and sustains the load

Hand trimmed trench bottom

Type B back fill

Selected excavated material free from large stones, roots of trees, vegetable matter, hardcore and other building rubbish and lumps of clay 75mm and above

FIGURE 113   (Class B bedding)

Normal backfill

Type B back fill placed and tamped by hand

Consolidated type B back fill in 100mm layers

Type A granular bedding well consolidated, and Y not less than 100mm

300(min)

Y

temporary drain

150mm (minimum)

Granular bedding material Type A.

Broken stone or gravel of 10mm size as specied in BSS 882

Other granular material and sizes up to 20mm may be suitable if readily compacted.

Anticipative problems	Considerations
1 **Load distribution on pipes**	1 Provide hand holes under each socket when using Class D bedding. This ensures that the pipe barrel rests on the trench bottom and not the collar, thus giving a larger, more uniform bearing surface for the pipeline. 2 Provide 100 mm of granular fill or 50 mm of lean concrete on bottoms of trenches excavated in rock. This removes the possibility of small rock projections causing breakages in drains when backfilled, resulting from point loads. 3 Ensure that where aggregate, especially in concrete haunching, has to flow under pipes that it is placed from one side only until it can be observed coming under the pipe on the other side. *Note* If the pipes are large and the fill is concrete, and it is placed on both sides at once, then air can be trapped beneath the pipe causing a void. See figure 115.
2 **Water seepage into workings**	1 Provide a land drain below and to one side of the proposed drain if not already considered. On completion plug the land drain with concrete 2 In peat and saturated clay soils seek local advice. This is essential if the clay lies over sand or gravel in which there is water which is subjected to artesian pressure because if this pressure is not reduced the trench bottom may heave and blow.
3 **Beddings near to but below existing foundation levels**	Provide a lean concrete backfill up to the level of the underside of the existing concrete foundations if the trench is 1 m or less, measured horizontally, from them. When the distance exceeds 1 m then the dimension measured vertically between the underside of the concrete foundations and the top of the concrete in the drain trench must be the horizontal measure less 150 mm. See figure 116.
4 **Gradients**	See figure 117. The gradient must be such as to provide a self cleansing velocity. It will vary with the pipe diameters, loading, and the number of houses connected to the drain.
5 **Runs generally**	1 Ensure that runs are straight between manholes and between levels. Bends are not allowed on the run but must be placed in the inspection chambers if and when required. 2 All junctions must occur in the manholes and must be generally in the direction of the flow. Three quarter channel sections are used in cases where the more expensive alternative of another inspection chamber would be unreasonable. 90° junctions should be avoided, and very short fittings with bends used with caution as some difficulty may be found in the insertion of the plugs for the hydraulic test. See figure 118.
6 **Pipes under buildings**	1 Use CI pipes, non-metallic rigid pipes with rigid joints surrounded with 150 mm of concrete, or flexible pipes surrounded with 150 mm of granular material are reinforcement in the concrete above to protect them from external loads and distortion. 2 Provide flexible joints within 600 mm of the wall face and a pea gravel surround not less than 150 mm thick, as illustrated in figure 119.

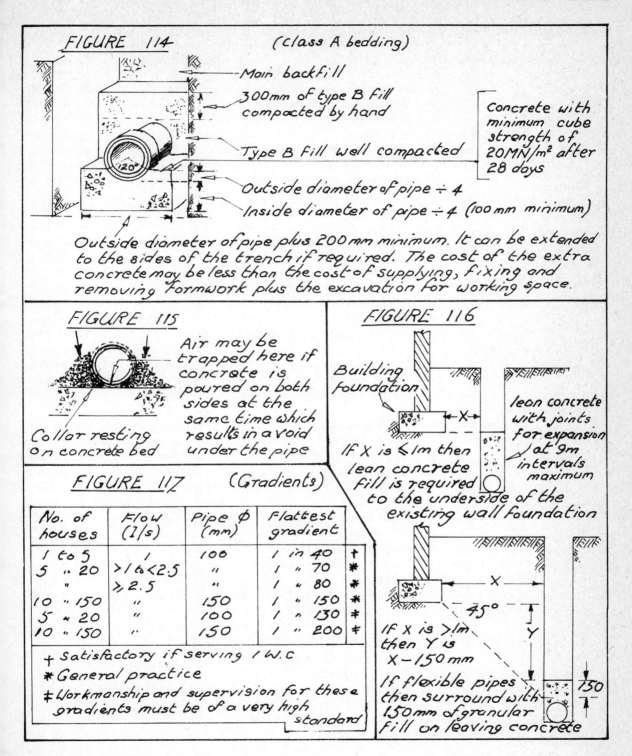

FIGURE 114 (class A bedding)

- Main backfill
- 300mm of type B fill compacted by hand
- Type B fill well compacted
- Outside diameter of pipe ÷ 4
- Inside diameter of pipe ÷ 4 (100mm minimum)

Concrete with minimum cube strength of 20MN/m² after 28 days

Outside diameter of pipe plus 200mm minimum. It can be extended to the sides of the trench if required. The cost of the extra concrete may be less than the cost of supplying, fixing and removing formwork plus the excavation for working space.

FIGURE 115

Air may be trapped here if concrete is poured on both sides at the same time which results in a void under the pipe

Collar resting on concrete bed

FIGURE 116

Building foundation

lean concrete with joints for expansion at 9m intervals maximum

IF X is ≤1m then lean concrete fill is required to the underside of the existing wall foundation

If X is >1m then Y is X−150mm

If flexible pipes then surround with 150mm of granular fill on leaving concrete

FIGURE 117 (Gradients)

No. of houses	Flow (1/s)	Pipe Ø (mm)	Flattest gradient	
1 to 5	1	100	1 in 40	†
5 " 20	>16<2.5	"	1 " 70	*
"	≥2.5	"	1 " 80	*
10 " 150	"	150	1 " 150	*
5 " 20	"	100	1 " 130	†
10 " 150	"	150	1 " 200	‡

† Satisfactory if serving 1 W.C
* General practice
‡ Workmanship and supervision for these gradients must be of a very high standard

Bedding for the pipes   *continued*

Anticipative problems	Considerations
	3 Provide a plate to cover the space between the pipe and the wall through which it passed. See also figure 119 and the next item about point loads on pipes.
7 **Point loads on pipes in manholes**	1 For shallow manholes provide a sand joint between the pipe and the wall over it. When the pipe is built over the sand may be removed and the joint left open until the wall is complete. Finally it can be filled with mastic on the completion of the work or, if no movement or settlement is anticipated, the same mortar as used for the manhole walls. 2 For pipes of 300 mm diameter and above then provide a rough relieving arch over the pipe and some provision for settlement. Where pipes have to pass through the walls of buildings sleeves may have to be built in, but in all cases allowing a 50 mm space for resilient packing together with a cover plate. See figure 119. 3 Provide flexible joints on either side of the wall as mentioned before. If it is thought that settlement might cause a sloping pipe to become flat or dip the other way then the initial fall might be slightly increased.
8 **Ground movement**	Provide flexible constructional joints in concrete beddings and haunchings at intervals of not more than 5 m. If flexible pipe joints have been used to cope with ground movement but the use of concrete beds is required for strength then the expansion joints in the concrete will give back to the drain some of the flexibility otherwise lost because of the need for concrete.
9 **Splitting of the pipe collars**	1 When flexible joints are used on a granular fill and the ground movement is expected to be large then place 'Denso' tape around the open end, or a band of soft clay or other suitable material. Small pebbles getting between the pipe socket and collar will restrict the required movement and may cause the collar to split. 2 Ensure that all hemp gaskets placed in a pipe joint are wet before filling with cement and sand. Dry gaskets will swell when wet, and if this occurs after the drain is put into service the expansion could cause the collar to fracture.
10 **Cover for pipes**	For some further illustrations on cover in deep and very shallow trenches see figure 120.
11 **Blockages and flow observations**	As shown in figure 123 provide inspection chambers on all drains to private properties at: 1 the head of the drain near the bend at base of the S & VP unless there is already a rodding eye at that point; 2 each change of direction on the pipe run; 3 every junction; 4 changes in levels. This will require a backdrop manhole. See figure 121; 5 90 m intervals (maximum) on straight runs; 6 a point within 12.5 m from the connection of the drain to another drain or sewer.

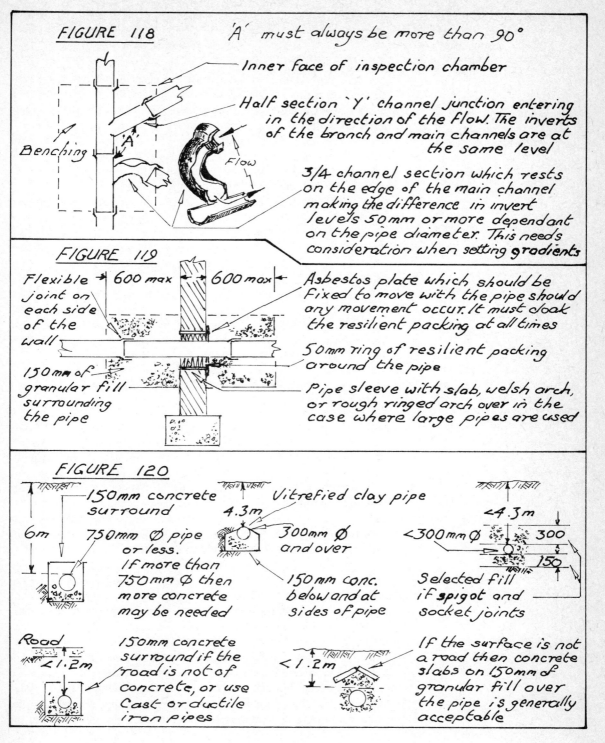

FIGURE 118

'A' must always be more than 90°

Inner face of inspection chamber

Half section 'Y' channel junction entering in the direction of the flow. The inverts of the branch and main channels are at the same level

Benching

'A'

Flow

3/4 channel section which rests on the edge of the main channel making the difference in invert levels 50mm or more dependant on the pipe diameter. This needs consideration when setting gradients

FIGURE 119

Flexible joint on each side of the wall

600 max   600 max

Asbestos plate which should be fixed to move with the pipe should any movement occur. It must cloak the resilient packing at all times

50mm ring of resilient packing around the pipe

150mm of granular fill surrounding the pipe

Pipe sleeve with slab, welsh arch, or rough ringed arch over in the case where large pipes are used

FIGURE 120

150mm concrete surround

Vitrefied clay pipe

6m

750mm Ø pipe or less. If more than 750mm Ø then more concrete may be needed

4.3m

300mm Ø and over

150mm conc. below and at sides of pipe

<4.3m

<300mm Ø

300

150

Selected fill if spigot and socket joints

Road

<1.2m

150mm concrete surround if the road is not of concrete, or use cast or ductile iron pipes

<1.2m

If the surface is not a road then concrete slabs on 150mm of granular fill over the pipe is generally acceptable

97

Bedding for the pipes *continued*

Anticipative problems	Considerations

Note   Access to the drains is essential in order that blockages can be removed from time to time. Also it may be necessary to trace the drain runs by observing the flow during a survey. Access for both of these things is essential and is provided by means of these inspection chambers but they can vary considerably in size, type, shape and depth, etc. See later notes on manholes.

12 Ventilation of the drains

1   Provide a S & VP at the head of each main drain run and at the head of each branch drain which is 6 m or more in length.

Note   Generally if each building has a soil and vent pipe then there is a constant flow of air through the drains keeping unpleasant smell to a minimum. Numerous S & V pipes on a single building can be costly, but often a diversion of the main run can eliminate the need for stacks other than the one at the head of the main run. See *Location of sinks* in *Sanitary Plumbing* and figure 110.

2   Provide a fresh air inlet at low level on the intercepting chamber on drains which are carrying sewage and waste to a cesspit or septic tank. It must be located so that a free passage of air can flow into the house drains, the trap being positioned on the sewer end of the manhole after any connection of the FAI to the drain. See figures 122 and 123.

13 Materials
(a) Rigid pipes

For resistance to chemical attack see CP 301:1971 table 4. Those in general use are saltglazed or vitrefied clay, concrete, asbestos cement and cast iron all with rigid or flexible joints.

1 Vitrefied clay pipes

Vitrefied clay pipes are superceeding the well tried saltglazed ware, and are suitable for most drains especially for acid soils and effuents. The pipes are longer than originally and this plus the patent push-fit joints reduce the main disadvantages which were the large number of connections required, and the time needed for them to cure. Clayware does not decompose or distort.

2 Concrete

This is suitable for ordinary ground but is susceptible to attack from acids or sulphates which are present in the effluent and surrounding soil. Pipes with ogee joints should not be used for carrying sewage as this type of joint cannot be guaranteed to stay watertight.

3 Asbestos cement

Whilst suitable for most domestic sewage it is susceptible to attack as is concrete but can be afforded protection as can concrete by coating in hot bitumen. Where doubt exists regarding the selection of these or concrete pipes because of the nature of the soil, seek the maker's advice.

4 Cast iron

This material is in common use and suitable for most drainage jobs. It is heavy but stronger than other rigid pipe materials, and resists corrosion better than wrought iron or steel. During handling, care is needed to ensure that the bituminous protective coating applied internally and externally during manufacture is not damaged, especially if it is to be

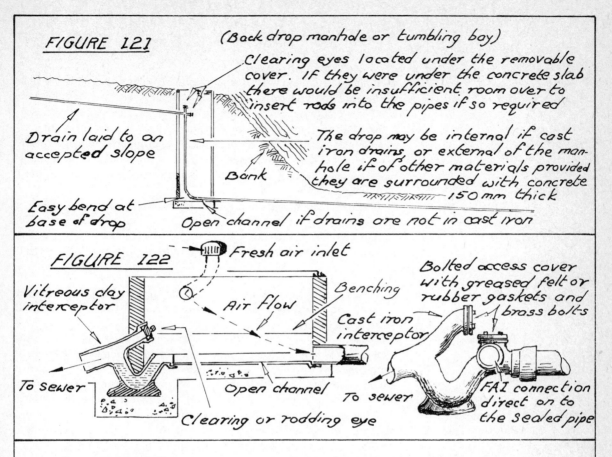

FIGURE 121 *(Back drop manhole or tumbling bay)*

Clearing eyes located under the removable cover. If they were under the concrete slab there would be insufficient room over to insert rods into the pipes if so required

Drain laid to an accepted slope

The drop may be internal if cast iron drains, or external of the man-hole if of other materials provided they are surrounded with concrete 150 mm thick

Bank

Easy bend at base of drop

Open channel if drains are not in cast iron

FIGURE 122

Fresh air inlet

Vitreous clay interceptor

Air Flow

Benching

Bolted access cover with greased felt or rubber gaskets and brass bolts

Cast iron interceptor

To sewer

Open channel

To sewer

Clearing or rodding eye

FAI connection direct on to the sealed pipe

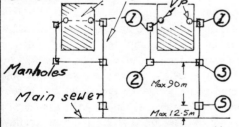

FIGURE 123

Branch drain ≥ 6m long which might be reduced by running the drain under the building as shown dotted

Manholes

Main sewer

VP

Max 90m

Max 12.5m

Note:- Air flows continually through the stacks, drains and sewer

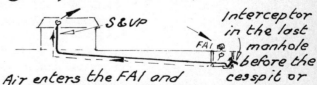

Manholes are required at
1. head of the drain if rodding eyes are omitted
2. bends
3. junctions
4. changes in level as shown in Figure 121
5. within 12.5m of the connection to the main sewer
6. at 90m intervals (max)

S&VP

FAI

Interceptor in the last manhole before the cesspit or septic tank

Air enters the FAI and leaves through the S&VP

Materials   *continued*

Anticipative problems	Considerations
	in contact with acid laden effluent or soil. Further protection is available in the form of acid resisting enamel or glass liners, or sheathing with 1000 gauge polythene or wrapping with water repellant tape externally. A 50% lap provides double protection. The pipes are brittle but provided tension is limited they may be used in shallow trenches, under roads and buildings, supported on piers or suspended from ceilings. Where impact damage may be anticipated then some protection must be provided. See fiugre 85.
(b) Flexible pipes	These pipes may be of ductile iron, pitch fibre or uPVC all of which may have rigid or flexible joints. The placing, packing and consolidation of the granular fill around pitch fibre and uPVC pipes requires expert care otherwise the loading above may cause distortion, blockages and leaks.
1 **Ductile iron**	This is expensive but has a higher resistance to corrosion than cast iron but has some of the qualities of steel in that it can be struck and dented without fracturing.
2 **Pitch fibre**	Pitch fibre is suitable for domestic waste, including some acids and alkalis at low temperatures. As it is not brittle it may be used where some uneven settlement is anticipated. The manufacturer's advice should be sought if the flow is to be continual and carry pitch solvents such as fats, oils, petrol or concentrations of acids. See figure 124.
3 **Unplasticised polyvinyl chloride**	uPVC is highly resistant to domestic effluents, many acids, alkalis and sulphates where the flow is intermittent. Where continual exposure is anticipated then advice on its use should be obtained from the manufacturers. This applies also when the waste is very hot and expected to flow for long periods, as can be expected from commercial kitchens and laundries. It is light in weight, easy to handle and obtainable in long lengths. Care is needed with the bedding of this material otherwise it may flatten, and the waste temperature must not exceed 65°C. See figure 125.
Flexible joints	These reduce laying time in that the curing of joints is not required, they are not affected by weather or wet conditions in the trenches, and testing can be done once the joint is completed. The use of lighter materials in longer lengths permit drains to be assembled quickly on ground level and lowered into the trench. This may also be possible with heavier pipes but a small crane would be employed and care is needed to avoid any collapse of the trench sides due to the weight or movement of it. Flexible joints allow the drains to follow shallow curves and provide for movement whereas rigid joints may result in shearing and leaks, and subsequent pollution of the ground.

FIGURE 124
(Pitch fibre drains)

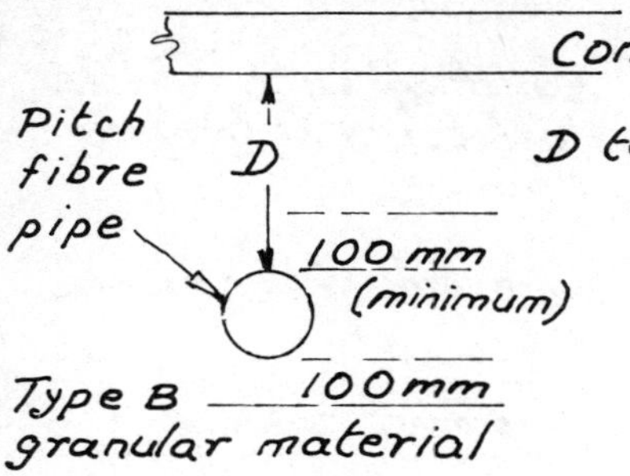

Concrete or tar macadem road

Pitch fibre pipe

D

100 mm (minimum)

Type B granular material

100 mm

D to be not less than 150mm of granular fill if under

(a) a concrete road

(b) the road formation level
or
600mm of the road surface whichever is the greater

if the surface is flexible such as tar macadem

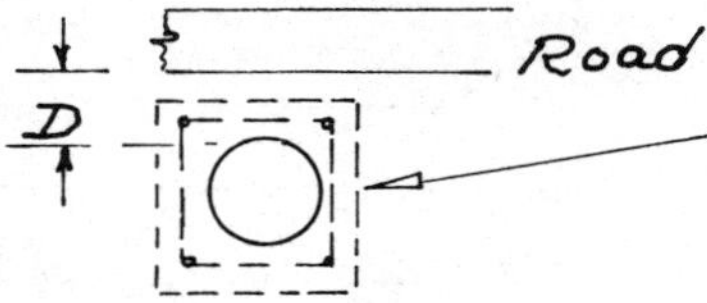

Road

D

150 mm concrete surround and reinforced if D is less than 150mm and the pipe is under a road with a flexible surface

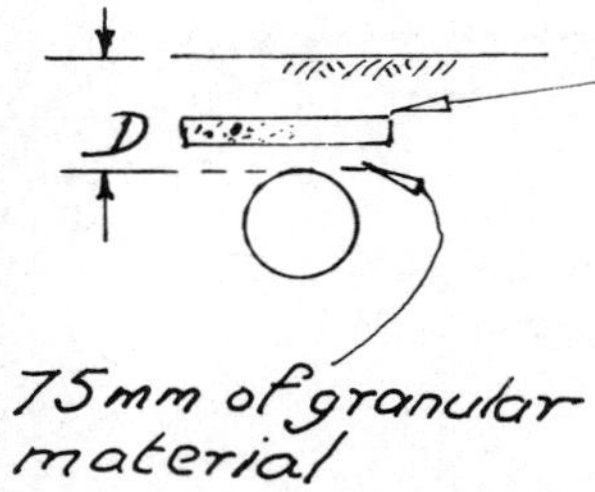

D

Concrete slabs over the pipe when the cover is less than 500mm and the pipe is not under a road

Maximum for D is 6m where the trench is narrow and in stable ground, and the water table is low. On sites where the water table is high then D would be limited to 3m

75mm of granular material

FIGURE 125

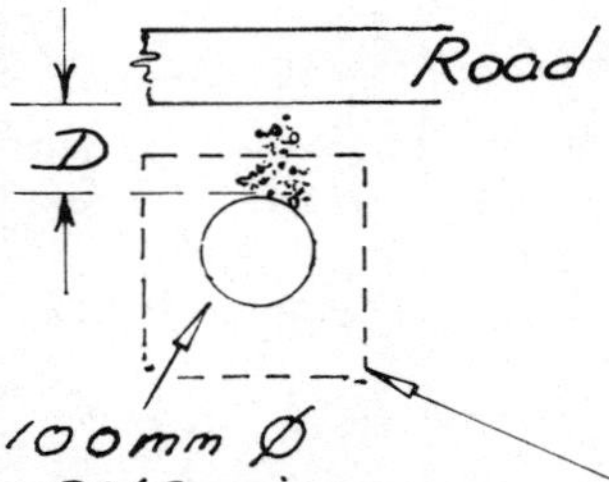

Road

D

1·2 m minimum under a light road unless surrounded with concrete

4·0 m maximum cover

D is $\Big\{$ 600mm minimum if under a road or hard standing

otherwise

as with shallow pitch fibre pipes

Using granular filling

100mm Ø uPVC pipe

Concrete surround required for locations under roads and where D is less than 1·2m

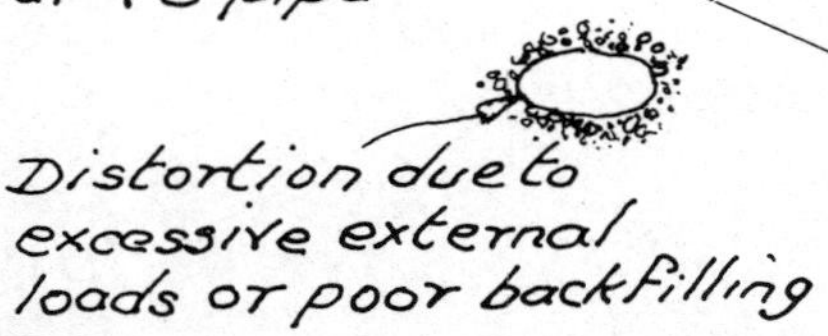

Distortion due to excessive external loads or poor backfilling

101

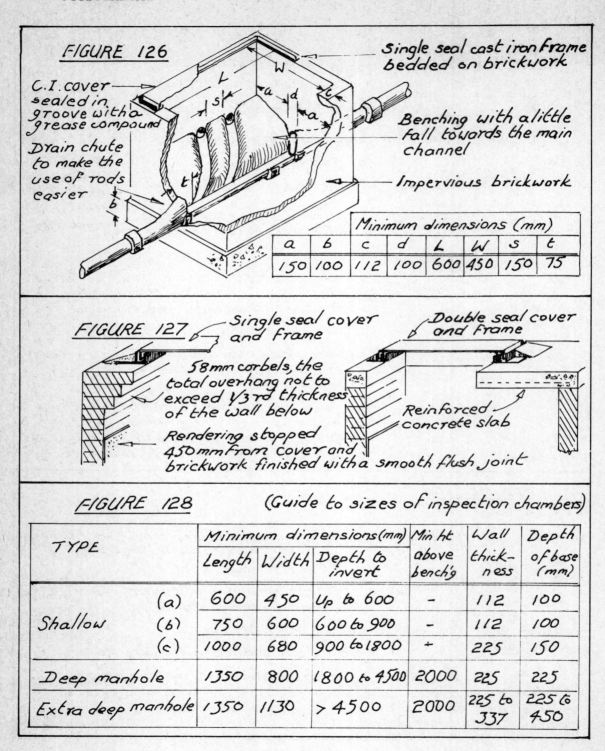

FIGURE 126

C.I. cover sealed in groove with a grease compound

Drain chute to make the use of rods easier

Single seal cast iron frame bedded on brickwork

Benching with a little fall towards the main channel

Impervious brickwork

Minimum dimensions (mm)							
a	b	c	d	L	W	s	t
150	100	112	100	600	450	150	75

FIGURE 127

Single seal cover and frame

Double seal cover and frame

58mm corbels, the total overhang not to exceed 1/3rd thickness of the wall below

Rendering stopped 450mm from cover and brickwork finished with a smooth flush joint

Reinforced concrete slab

FIGURE 128   (Guide to sizes of inspection chambers)

TYPE		Minimum dimensions (mm)			Min ht above bench'g	Wall thickness	Depth of base (mm)
		Length	Width	Depth to invert			
Shallow	(a)	600	450	Up to 600	-	112	100
	(b)	750	600	600 to 900	-	112	100
	(c)	1000	680	900 to 1800	+	225	150
Deep manhole		1350	800	1800 to 4500	2000	225	225
Extra deep manhole		1350	1130	> 4500	2000	225 to 337	225 to 450

Manholes and inspection chambers

Inspection chambers are usually shallow, and on being opened up permit one to check on drain runs or insert drain rods for removing blockages from the pipes, both of which can be done from ground level. Manholes are really deep inspection chambers, built for the same reason but as they are deep they have to be larger in order that a man can not only get down into it, but can actually insert the clearing equipment at that low level if so required.

Anticipative problems	Considerations
1 Location of the manholes	See previous problem 11 'Blockages of pipes' and figure 123.

2 Size of the units	1 Dimensions of the traditional brick inspection chambers with the channel built *in situ*, as illustrated in figure 126 are:
	    (a) minimum internal width is the diameter of the channel plus not less than 150 mm each side for the benching;
	    (b) minimum length is dependant on the diameter of the branch pipes and the number of them on any one side. Allow also not less than 150 mm between branches and 75 mm between the end branch and wall. *Note* Three branches on any one side is usually the maximum as the length at invert level will otherwise exceed the cover length. To reduce the chamber plan dimensions to cover size then either
	        (i) corbell the walls over at the top or
	        (ii) provide a reinforced concrete slab with a perforation for access. See figure 127.
	    (c) depth will depend on the diameter of the main flow pipe, the amount of fall on the run and the ground contour at that point, the chamber being classified as in figure 128.
	Note Some modern type manholes in concrete and glass fibre are much more convenient and time saving but their depths need to be limited to about 1 m otherwise some difficulty may be experienced in inserting clearing rods for the removal of blockages. See figure 133 remembering that the deeper the manhole the more difficult it becomes to manœuvre the rods into pipes.
	    Very deep traditional type manholes will require an access shaft not less than 600 x 650 mm internally carried on a reinforced concrete slab. See figure 130.

3 Type of unit required	1 **Normal inspection chambers**. These may be factory made units in concrete, glass fibre, or plastics or the traditional brick unit built *in situ*. They are used at the head of the drain, within 12.5 m of the connection to the sewer, at all changes of direction, junctions and at not more than 90 m intervals on the run. See figures 123 and 126. 2 **Intercepting chambers** These are not as common as they used to be, but when the local authority does require them they may be (a) the last manhole on a foul drain run located just before discharging into a cesspool, septic tank or public sewer. The water seal in the interceptor trap prevents obnoxious smells and gases from the sewer entering the house drains. See figures 122 and 129.

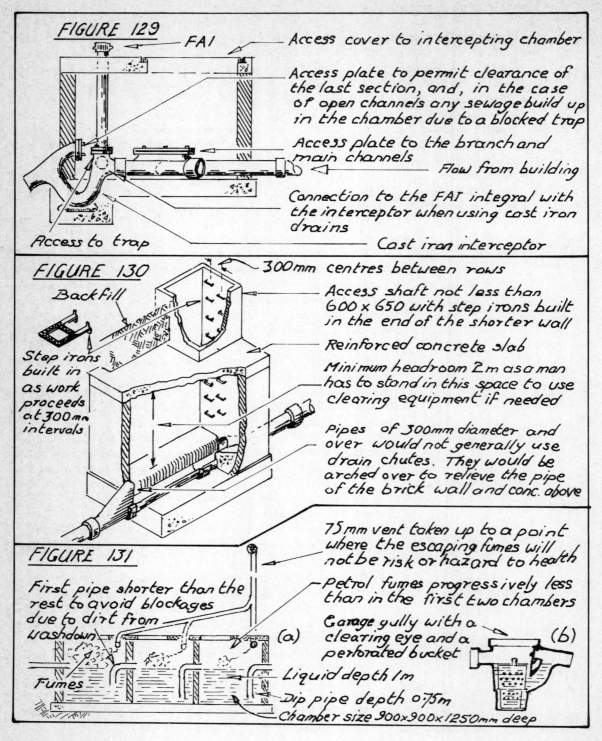

FIGURE 129

FAI

Access cover to intercepting chamber

Access plate to permit clearance of the last section, and, in the case of open channels any sewage build up in the chamber due to a blocked trap

Access plate to the branch and main channels

Flow from building

Connection to the FAI integral with the interceptor when using cast iron drains

Cast iron interceptor

Access to trap

FIGURE 130

Backfill

Step irons built in as work proceeds at 300mm intervals

300mm centres between rows

Access shaft not less than 600 x 650 with step irons built in the end of the shorter wall

Reinforced concrete slab

Minimum headroom 2m as a man has to stand in this space to use clearing equipment if needed

Pipes of 300mm diameter and over would not generally use drain chutes. They would be arched over to relieve the pipe of the brick wall and conc. above

FIGURE 131

First pipe shorter than the rest to avoid blockages due to dirt from washdown

Fumes

75mm vent taken up to a point where the escaping fumes will not be risk or hazard to health

Petrol fumes progressively less than in the first two chambers

Garage gully with a clearing eye and a perforated bucket

(a)

(b)

Liquid depth 1m

Dip pipe depth 0.75m

Chamber size 900x900x1250mm deep

Manholes and inspection chambers *continued*

Anticipative problems	Considerations

(b) the inspection chamber where the storm water drain using shoes at the bottom of each rain water pipe finally joins the foul drain. This type is a reverse interceptor and is located at the inlet to the chamber and not as in (a) where it becomes the outlet. See figure 135.

Note   The water seal in this case prevents the rain water pipes acting as vent pipes for the sewer and drains and discharging the gases and smells at eaves level near open windows.

3 Drop manholes   These are essential where the invert level suddenly charges, caused by a variation in ground level such as a bank, or a flight of steps (See figure 121.

4 Deep manholes   See figure 130. For these provide

(a) an access shaft not less than 600 x 650 mm measured on plan, located on one side or in one corner of the chamber;

(b) a reinforced concrete cover slab to carry the shaft and the earth backfill.

Note   Whilst the whole area on plan has to be excavated down to the foundation level, the actual chamber itself needs only to be high enough to permit a man to stand up and work. It is not necessary or economic to take it higher, hence the backfill of earth over the slab.

(c) galvanised step irons built in as the work proceeds, these being fitted on the short wall of the shaft and continuing down into the main chamber. If a ladder is built in then the length of the shaft, on plan, needs to be extended for at least another 150 mm.

5 Petrol interceptors   Generally the local Petroleum Officer will be in control concerning any risks arising from petrol and oil spillages. Basically it could be as follows:

(a) No special precautions required for one to nine cars inclusive. However if gulleys are to be provided for washdown facilities then for single and two car units where the gulley is placed externally it should be of the deep seal type otherwise evaporation of the seal during dry periods and the subsequent smells could be a constant worry and health danger for the occupants.

(b) Garage gulleys may be stipulated for ten to fortynine parked cars. See figure 131 (b).

(c) Provide a petrol interceptor when the number of cars is fifty or over. See figure 131 (a).

The requirements will vary with the anticipated risk and some authorities may demand an interceptor for under 50 vehicles but possibly accept one with only two instead of three chambers.

Note   Where car parks are proposed for spaces below ground then another important aspect to consider is ventilation because of the possible pockets of concentrated carbon monoxide. Other considerations for these units are to:

1 ensure that the vent pipe from the chambers terminates externally and is not less than 2.4 m above ground level, or, if windows are nearby, that the terminal is always above them and by not less than 1 m when less than 3 m from them.

Manholes and inspection chambers *continued*

Anticipative problems	Considerations

<table>
</table>

4 Type of cover

2 ensure that the reinforced concrete slab and cover is adequate to carry the anticipated wheel loads from above.

These are available with single or double seals and may be rectangular, square made up of two triangular sections, triangular or circular. Generally they fall into the following classes.

(a) Heavy duty for use in carriage ways where wheel loads between 11 and 12 tonnes can be anticipated. Ventilating types are available if required.

(b) Medium duty for use in public footpaths, access to domestic properties, verges and cycle tracks. These are usually all single seal types and either circular or rectangular in shape.

(c) Light duty for areas which are inaccessible to wheeled vehicles. They are rectangular and may be single or double sealed.

Double sealed units only must be used internally and they must be fixed down with brass screws. Where covers have to be exposed internally then recessed types are available which can be filled with the same finishes as the floor, possibly with brass edges to the frame and cover where cast iron and rust might otherwise be a problem.

5 Essential construction details

1 Avoid rendering the top six courses of brickwork internally, especially if the bricks are porous and the water table is high, otherwise in winter the saturated wall can freeze and the subsequent expansion caused by the ice formation may push rendering and even parts of bricks themselves away. If these fall into the channel then it will cause a blockage of the pipes. It is better to use non-porous bricks without rendering, and finish the brickwork with smooth flush or weather struck joints. See figure 127.

2 If step irons are to be used ensure that the access cover and frame or access shaft have one or two walls which are or the continuation of the chamber walls, that is on one side or in one corner. See figure 130. *Note* If the shaft or cover is over the centre of the chamber then the only way one can get in or out of it is by the use of a ladder which is not practical. The step irons could be inaccessible to a workman suspended in the central access hole.

3 Ensure that the benching extends for not less than 150 mm above the top edge of the channel and that it slopes towards it, otherwise should any sewage be spilled on to the upper surface it can remain fouled indefinitely.

4 Provide chains on clearing eyes of interceptors and attach the top end to a ring fixed just below the cover to ensure that it is easily accessible should the need arise. See figure 132.

Note Should a blockage occur in the interceptor this chamber will fill, but the only indication of a fault will be the rising and falling of the water in the pan during each flush. On lifting the cover the clearing eye, which has to be removed in order that the contents flow away, will not be visible as it will be submerged in the sewage. If no chain exists then

FIGURE 132

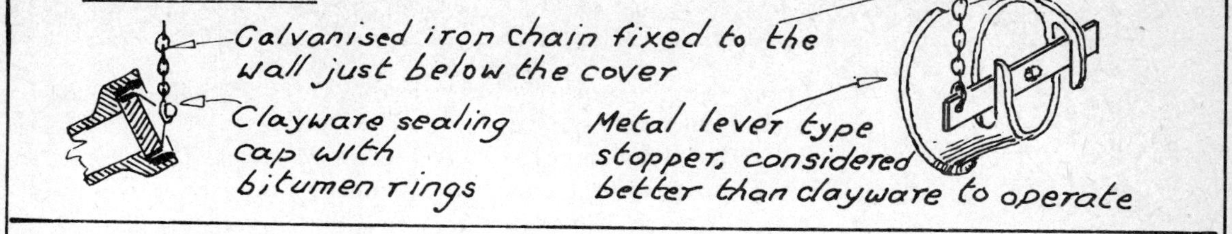

Galvanised iron chain fixed to the wall just below the cover

Clayware sealing cap with bitumen rings

Metal lever type stopper, considered better than clayware to operate

FIGURE 133

CI Frame

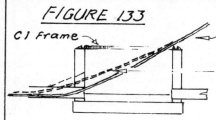

The rods engage with the top edge of the pipes, especially if the flexibility is poor

The dotted line illustrates the extra free space offered by the drain chute (shown dotted) which reduces the wear and tear on the cleaning equipment

FIGURE 134

RWP

Inlets in the grid for waste pipes if needed

Inlets may be vertical or horizontal and located on the back or side of the gully

Trapped gully at the base of each rainwater pipe

Trap may be of the 'P' or 'S' pattern, and in some gullies a connection, shown dotted, is provided and fitted with a removable cap to permit easy clearance of the drain

FIGURE 135

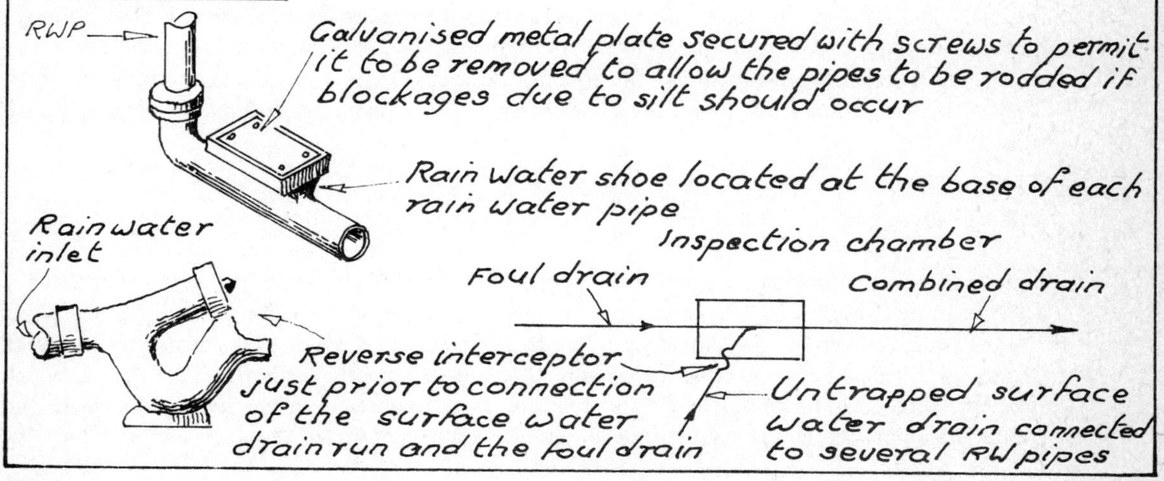

RWP

Galvanised metal plate secured with screws to permit it to be removed to allow the pipes to be rodded if blockages due to silt should occur

Rain water shoe located at the base of each rain water pipe

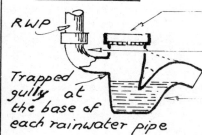

Rainwater inlet

Inspection chamber

Foul drain

Combined drain

Reverse interceptor just prior to connection of the surface water drain run and the foul drain

Untrapped surface water drain connected to several RW pipes

Manholes and inspection chambers *continued*

Anticipative problems	Considerations

to say that the removal of this clearing eye is an unpleasant task is a considerable under-statement.

5 Provide drain chutes in all inspection chambers. This permits any clearing rods to be pushed into the drains and removed more easily, especially if the chamber is other than shallow. See figures 126 and 133.

Combined drains

Combined drains are used on sites completely occupied by the building and where storm water sewers are not available and soakaways are not possible. Because storm water and sewage both run in the same pipe eventually it is possible for the rainwater pipes to operate as vent pipes for the sewer, but as the RWP must terminate at the gutter which could be near openable windows this then becomes a health hazard and would not comply with Building Regulations. Just as sanitary units are isolated from the drains by the use of traps and water seals then so must the rainwater pipes, and it can be done in two ways by either

1   discharging each rain water pipe into a trapped gulley before joining the foul drain. See figure 134;

2   connecting the rain water pipes through rain waters shoes to a surface water drain which in turn is fitted with a reverse interceptor at the inspection chamber just prior to it joining the foul drain. See figures 135, 136 and 149.

Note   Not all authorities will permit the use of combined drains as storms can cause problems at the treatment works, especially if the overflow tanks are not adequate to carry the extra volume of polluted water. It is advisable to check with them early on to see what is required and what facilities are available before deciding on the type of drain to be adopted.

Also the separate foul and surface water drains usually involve the crossing over of pipes until the combined drain is established. This will mean extra excavation and back filling for some runs to allow these passings. Levels should be checked to determine whether or not it is possible especially as both have to meet the same level when they combine. It is possible that although surface water may have a flatter gradient than the foul ones, the initial starting levels may need to be appreciably different. If the foul drains in figure 136 ran under the building cross-overs would be avoided but it has the disadvantages of the manholes and maintenance being all internal.

Raising sewage to a higher level

Sometimes it happens that the appliances are below the level of the drain so that in order that the sewage can be disposed of it has first to be raised to a higher level. Although facilities for doing this are available every effort should be taken to avoid the need for it as it constitutes the weak link in the chain. Should it cease to operate then unless standby equipment is installed, which obviously increases the initial costs, and requires extra maintenance and more floor space, no sewage can be moved, and in a short time all sanitary appliances would be out of use.

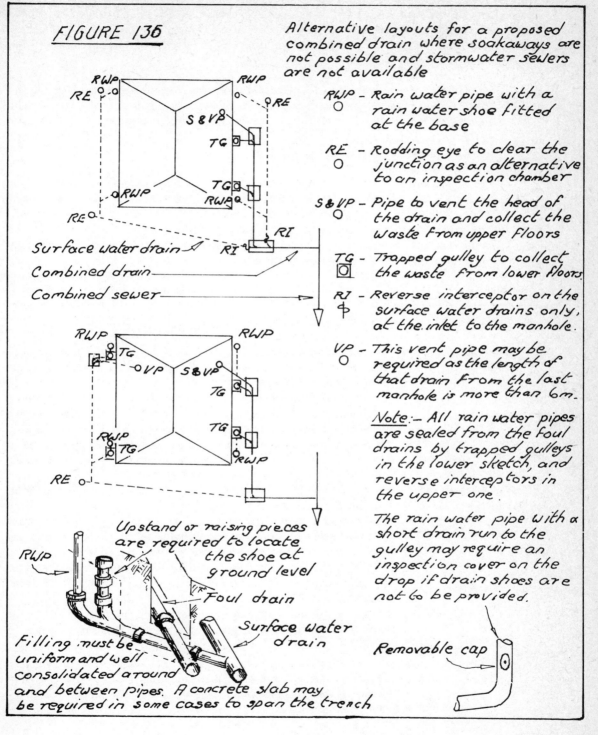

FIGURE 136

Alternative layouts for a proposed combined drain where soakaways are not possible and stormwater sewers are not available

RWP - Rain water pipe with a rain water shoe fitted at the base

RE - Rodding eye to clear the junction as an alternative to an inspection chamber

S&VP - Pipe to vent the head of the drain and collect the waste from upper floors

TG - Trapped gulley to collect the waste from lower floors

RI - Reverse interceptor on the surface water drains only, at the inlet to the manhole.

VP - This vent pipe may be required as the length of that drain from the last manhole is more than 6m.

Note:- All rain water pipes are sealed from the foul drains by trapped gulleys in the lower sketch, and reverse interceptors in the upper one.

The rain water pipe with a short drain run to the gulley may require an inspection cover on the drop if drain shoes are not to be provided.

Surface water drain
Combined drain
Combined sewer

Upstand or raising pieces are required to locate the shoe at ground level

RWP
Foul drain
Surface water drain

Filling must be uniform and well consolidated around and between pipes. A concrete slab may be required in some cases to span the trench

Removable cap

109

The sewage can be raised directly for a short distance to a level where gravity can once again control any further flows, or it may have to be pumped up hill which would require considerably more power. Also provision must be made to prevent any fluid which has passed through the lifting plant from flowing back into it once the lifting power ceases to operate. The type of plant usually employed for this work is

(a) *Pneumatic ejectors*

These are not recommended for large installations where the loads are considerable and variable, but more for situations such as isolated sites where reliability and ease of maintenance takes priority over efficiency. As the lifting force is compressed air then these ejectors have an advantage over pumps in that several may be operated from a single central compressor or pressure cylinder if not too far away.

(b) *Sewage lifting pumps*

These are advisable for the larger installations where the flows are continuous but may be variable. The mechanical plant can be designed to cope with maximum flows passing through minimum sized pipes and all the variations which can occur. As the loads may be high the pump house needs to be located away from the building in order to avoid and noise or smells which could be a source of annoyance. It needs to be dry, and easily accessible to electrical supplies and for maintenance.

Anticipative problems	Considerations
	Pneumatic ejectors and compressed air units See figure 137
1 Noise and the location	1 The compressor may be located with the unit or separate from it. The exhaust needs to be well silenced or located at a point where the noise is of no consequence.
	2 The number of times the compressor will operate may be reduced considerably by the use of a compressed air cylinder which will provide the motive power for the ejector to operate several times once fully charged by the compressor. Extra space will obviously be required to house this pressure cylinder and it is the usual practice to duplicate this unit if a standby compressor is to be installed.
2 Operation of the unit	Provide a sump for the collection of the sewage and a separate space for for the float chamber. As designs can vary then consult the manufacturers regarding their specific needs, for example, if air pumps have to be used then bore holes of some considerable depth have to be made in order that the pipes are submerged for the correct depth. *Note* The rising of the float caused by the inflow of sewage from the collector chamber will open a valve and release compressed air to the float chamber. This will prevent any further inflow but will force the sewage out through the delivery pipe causing the float to fall as the level drops. When the float reaches the limit of its fall it will close the air valve allowing fresh sewage to flow again into the float chamber.
3 Breakdowns	Provide twin compressors and, if used, two compressed air cylinders. This is recommended in order that should one fall then the removal of sewage is still possible whilst spare parts are being obtained and the faulty unit is being repaired.

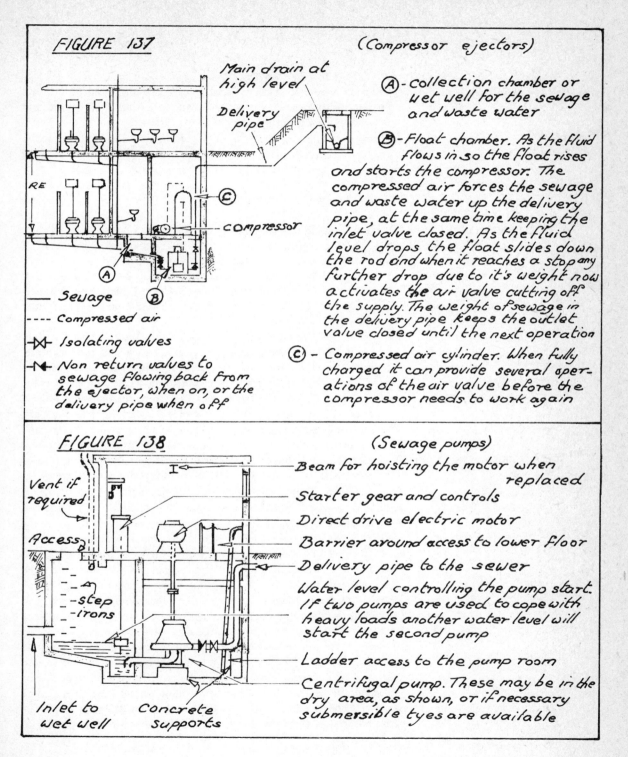

FIGURE 137 (Compressor ejectors)

Main drain at high level

Delivery pipe

(A) - Collection chamber or wet well for the sewage and waste water

(B) - Float chamber. As the fluid flows in so the float rises and starts the compressor. The compressed air forces the sewage and waste water up the delivery pipe, at the same time keeping the inlet valve closed. As the fluid level drops the float slides down the rod and when it reaches a stop any further drop due to it's weight now activates the air valve cutting off the supply. The weight of sewage in the delivery pipe keeps the outlet valve closed until the next operation

(C) - Compressed air cylinder. When fully charged it can provide several operations of the air valve before the compressor needs to work again

— compressor

RE

—— Sewage
---- Compressed air
Isolating valves
Non return valves to sewage flowing back from the ejector, when on, or the delivery pipe when off

FIGURE 138 (Sewage pumps)

Vent if required

Access

Beam for hoisting the motor when replaced
Starter gear and controls
Direct drive electric motor
Barrier around access to lower floor
Delivery pipe to the sewer
Water level controlling the pump start. If two pumps are used to cope with heavy loads another water level will start the second pump
Ladder access to the pump room
Centrifugal pump. These may be in the dry area, as shown, or if necessary submersible tyes are available

step irons

Inlet to wet well
Concrete supports

Pumps and ejectors   *continued*

Anticipative problems	Considerations
4 Maintenance	**1** Provide adequate and easy access to all chambers and ensure that doors, access traps etc, are large enough to permit the removal and replacement of faulty units. **2** Provide lifting beams above units, (see figute 138) or above access holes through which plant has to be hoisted during periods of some maintenance or should it have to be replaced.
	Sewage lift pumps. See figure 138

1 Location of the unit	Provide a pumping station which must be
	   (a) above or below ground level, but owing to the need for electrical supplies it is essential for the space to be kept completely dry.
	   (b) away from buildings if possible as it may be dealing with considerable quantities of sewage, and noise and smell might be a nuisance.
	   (c) adjacent to the collection chamber for the sewage if float operated pumps are used.

2 Operation of the pump	The installation should be to the manufacturer's instructions. The float rises on the flow of the incoming sewage and when it reaches a pre-determined level it operates an electrical control switching on the pumps. The sewage is physically forced into the delivery pipe and as the level in the collection chamber or wet well drops the float falls eventually actuating that same electrical control and switching the pumps off. When twin pumps are installed, one giving assistance when the demand is high, then it will have a different water level to bring it into use.
3 Failure of the pump	As mentioned under pneumatic ejectors this eventually must be covered and the method used may vary between the following possibilities **1** using twin pumps, each capable of coping with the full load if so required, but one kept in reserve as a standby should the other one fail; **2** using twin pumps, each capable of dealing with two thirds of the load, the second pump coming into operation when the load is too much for the single pump and when the usual pump for that period breaks down. This is obviously cheaper initially as the pumps are rated lower powerwise, but the fact that any single unit cannot cope with all loads could cause occasional problems. The decision on the choice must depend upon the money that is available at the time as against the need to cope with the odd periods when the load might be excessive. **3** using three pumps each able to cope with two thirds of the anticipated highest load. One would operate for normal flows and the second one would be brought into use when the quantity needed to be moved exceeded this general flow. The third pump would then be kept as a standby to be used only when one of the others failed. Whilst this is the most expensive, the flow of sewage should never be stopped due to faulty pumps.

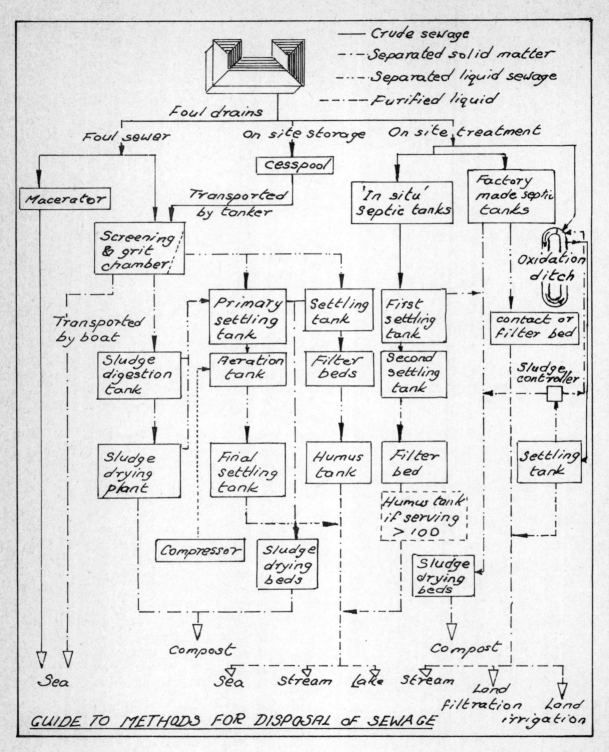

Legend:
——— Crude sewage
—·—· Separated solid matter
—··— Separated liquid sewage
– – – Purified liquid

Foul drains

Foul sewer | On site storage | On site treatment

Macerator

Cesspool

Transported by tanker

'In situ' septic tanks | Factory made septic tanks

Screening & grit chamber

Oxidation ditch

Transported by boat

Primary settling tank | Settling tank | First settling tank | contact or filter bed

Sludge digestion tank | Aeration tank | Filter beds | Second settling tank

Sludge controller

Sludge drying plant | Final settling tank | Humus tank | Filter bed | Settling tank

Humus tank if serving >100

Compressor | Sludge drying beds

Sludge drying beds

Compost

Compost

Sea | Sea | Stream | Lake | Stream | Land filtration | Land irrigation

GUIDE TO METHODS FOR DISPOSAL OF SEWAGE

5 Sewage disposal

In most cases the removal of sewage is no problem as it is taken away in the drains to the Local Authority's plant to be treated or disposed of direct. For some rural sites and isolated buildings the main drainage systems are non-existant so either the sewage has to be stored and collected periodically, or treated on site before disposal into streams or dispersal over the ground by irrigation or filtration.

Cesspits

Cesspits are provided on isolated sites where main drainage is non-existant, larger units such as septic tank installations are not possible because of restricted site space, and there is no means of disposing of the effluent on the land available. No treatment is given to the sewage and it is removed in a tanker periodically, the interval between visits depending on the number of users and the capacity.

Anticipative problems	Considerations
1 Actual capacity	Ensure that the capacity is between 18000 litres and 45000, measured below the inlet. Allow 136 litres per person and 2 to 8 persons per unit. COP 302.200 recommends a minimum storage time of 45 days. More frequent emptying generally requires for special arrangements to be made, plus additional expenses.
2 Dimensions of the cesspit	The maximum depth of the stored sewage and waste should not exceed 4 m. The depth measured from invert level of the inlet to the ground level should be between 750 mm and 1 m. To calculate the other dimensions see figure 140.
3 Location so as to avoid health hazards	1 On sloping sites ensure that the cesspit is located in ground at a lower level than the building it serves. This also helps towards an economic installation as the drain can follow the contours. See figure 139: 2 Ensure that any water source such as wells or springs are at a higher level than the cesspit. 3 Ensure that the distance of the pit from the nearest building complies with the Local Authortiy's requirements and is not a health hazard. 4 Check the levels of sanitary appliances in any proposed basements to ascertain whether any drainage from them is possible without pumping.
4 Construction requirements	1 Ensure that the walls and floor are impervious otherwise the effluent may leak out and polute the surrounding soil, or ground water may seep in and reduce the volume of sewage able to be stored. 2 Provide a reinforced concrete cover with a hole to provide access for the removal of sewage.

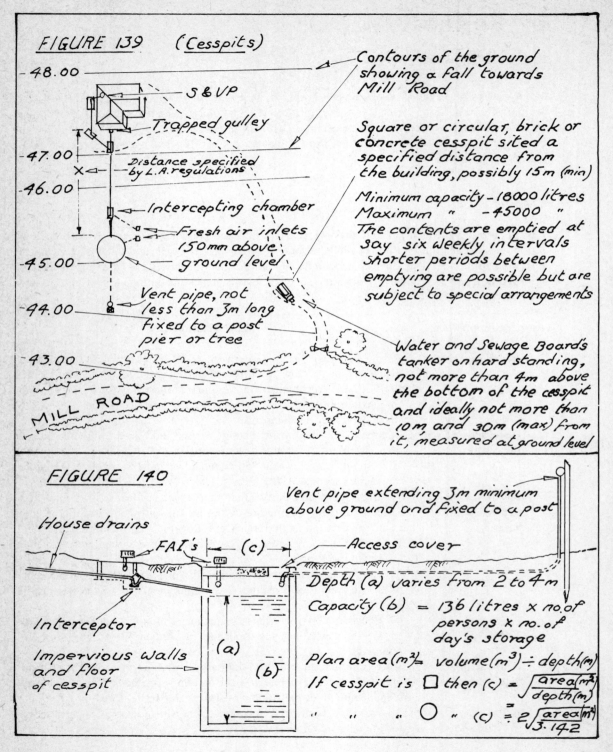

FIGURE 139 (Cesspits)

-48.00

S & VP

Trapped gulley

-47.00

Distance specified
by L.A. regulations

X

-46.00

Intercepting chamber

Fresh air inlets
150mm above
ground level

-45.00

Vent pipe, not
less than 3m long
fixed to a post
pier or tree

-44.00

-43.00

MILL ROAD

Contours of the ground
showing a fall towards
Mill Road

Square or circular, brick or
concrete cesspit sited a
specified distance from
the building, possibly 15m (min)

Minimum capacity - 18000 litres
Maximum " - 45000 "
The contents are emptied at
say six weekly intervals
shorter periods between
emptying are possible but are
subject to special arrangements

Water and Sewage Board's
tanker on hard standing,
not more than 4m above
the bottom of the cesspit
and ideally not more than
10m and 30m (max) from
it, measured at ground level

FIGURE 140

House drains

FAI's

(c)

Interceptor

Impervious walls
and floor
of cesspit

(a)

(b)

Vent pipe extending 3m minimum
above ground and fixed to a post

Access cover

Depth (a) varies from 2 to 4m

Capacity (b) = 136 litres x no.of
persons x no.of
day's storage

Plan area (m²) = volume (m³) ÷ depth (m)

If cesspit is ☐ then (c) = $\sqrt{\dfrac{area(m^2)}{depth(m)}}$

" " " ◯ " (c) = $2\sqrt{\dfrac{area(m^2)}{3.142}}$

116

Cesspits *continued*

Anticipative problems	Considerations

Note Avoid the use of built in step irons as portable ladders, if required, are more acceptable.

3 Provide an interceptor on the inlet side of the cesspit and the necessary FAI. See figure 139.

4 Provide adequate ventilation for the cesspit. A FAI is required 150 mm above ground level and a high level vent supported on a post, pier, tree, etc, and terminating not less than 3 m above the ground level. See figure 140.

5 Provide a hard standing for the tanker when emptying the pit. This must not be more than 30 m from the pit and not more than 4 m above the level of the base of it. See figure 139.

6 Ensure that no cesspit has an overflow. It is possible that one might be permitted if connected directly to a reserve pit installed to cover unforseen circumstances such as roads blocked with snow preventing visits by the tanker when needed.

5 Special precast units	Check with the manufacturer on the installation requirements when the soil is not free draining and the water table rises to less than 2 m below the surface of the ground.

Note Special precautions may be necessary to ensure that the unit does not float out of the ground during and after the installation.

Septic tanks

Septic tank installations provide biological treatment to the sewage and ensure that the polluting factors are removed from the liquid or rendered harmless before it is discharged over the site or into an adjacent stream. It comprises settling tanks in which anaerobic bacteria break down the solids into liquid form, and contact or filter beds where aerobic bacteria remove the solids left in suspension before finally disposing of the liquid. Sludge settles in the bottom of the septic tanks and once or twice a year this is removed and either taken away in a tanker or placed in beds over the site and dried before being used as compost. This plant is designed for large isolated structures and groups of small buildings located on sites which have no main drainage available, but have room to dispose of the treated effluent. The type of soil on site must permit it to disperse evenly and harmlessly when other outlets such as streams are non-existant.

Anticipative problems	Considerations
1 Capacity of the units	See figure 141. P in the formula is the total number of full-time and half the number of part-time building occupants.
2 Dimensions of the septic tank	The length is usually three times the width if a single tank is used. When two tanks are demanded then this length may be split using a

117

Septic tanks _continued_

Anticipative problems	Considerations

2:1 ratio. The depth below the invert is 1.5 to 1.8 m. Area of the single tank is

$$\frac{\text{Volume (m}^3\text{)}}{\text{Depth (m)}}\,\text{m}^2 \text{ so the width (W) is } \frac{\text{Area (m}^2\text{)}}{3}$$

Length of a single tank is 3 W, otherwise 2 W and W if two tanks are installed. See figures 141 and 142.

3 Location of the tanks

1 Ensure that the soil will permit the drainage of the treated effluent, for example, a septic tank installation would be unsuitable if

(a) the ground is heavy clay, unless the required standard of purification could be obtained without the need for land filtration and it could be fed direct to a stream.

(b) the water table is likely to be within 1.5 m of ground level for periods of several weeks.

Note In some cases where the land is suitable but too small in area at low level for filtration the Local Authority may permit the clarified efflent to be pumped to a higher level before dispersal over or into the ground.

2 Check the minimum distance the plant may be sited from the nearest building. This will vary with the number of users. See figure 141.

3 On sloping sites locate the plant on lower levels than the building it serves, and allow space between the filter and the boundary for the disposal of the treated effluent. See figure 141 and the _Note_ in 1 above.

4 Ensure that the plant is away from trees, especially if deciduous, as the falling leaves may block the serrated channels and filters unless protected with wire mesh or other material which will keep them out but not exclude the air.

4 Construction requirements

1 All units must be impervious.

2 Floors in the settling tanks must fall towards the inlet end which in turn forms a sump for the sludge. The floors to the filter must fall towards the outlet. See figure 142.

3 Provide biological filters and protect the media from falling leaves with wire mesh or other means. For additional data see figure 141.

Note Dosing chambers are often provided just prior to the filter to control the amount passed over the channels and aggregate each time.

4 Provide a humus tank for all installations where the number using it exceeds 100. With some circular filters the humus tank may be in the form of several small ones, one being located at each outlet around the base of the filter bed. For sizing data see figure 141.

5 Provide an area for gravity irrigation or land filtration if the discharge is not direct to a stream from a humus tank. To calculate the areas required see figure 141. The ground allocation must be duplicated

118

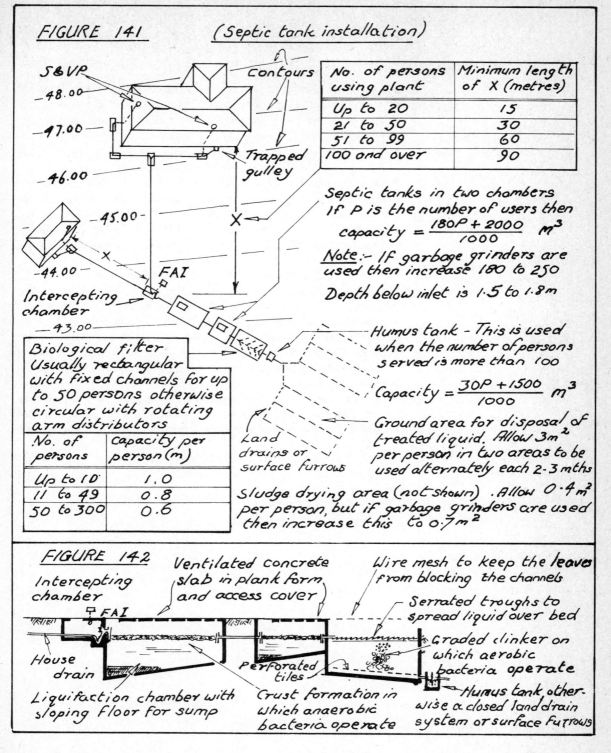

FIGURE 141   (Septic tank installation)

S&VP

−48.00

−47.00

−46.00

−45.00

−44.00

−43.00

Contours

Trapped gulley

X

FAI

Intercepting chamber

No. of persons using plant	Minimum length of X (metres)
Up to 20	15
21 to 50	30
51 to 99	60
100 and over	90

Septic tanks in two chambers
If P is the number of users then

$$capacity = \frac{180P + 2000}{1000} \ m^3$$

Note:- If garbage grinders are used then increase 180 to 250

Depth below inlet is 1·5 to 1·8m

Humus tank - This is used when the number of persons served is more than 100

$$Capacity = \frac{30P + 1500}{1000} \ m^3$$

Ground area for disposal of treated liquid. Allow 3m² per person in two areas to be used alternately each 2-3 mths

Land drains or surface furrows

Sludge drying area (not shown). Allow 0·4 m² per person, but if garbage grinders are used then increase this to 0·7m²

Biological filter
Usually rectangular with fixed channels for up to 50 persons otherwise circular with rotating arm distributors

No. of persons	Capacity per person (m)
Up to 10	1.0
11 to 49	0.8
50 to 300	0.6

FIGURE 142

Intercepting chamber

FAI

House drain

Liquifaction chamber with sloping floor for sump

Ventilated concrete slab in plank form and access cover

Perforated tiles

Crust formation in which anaerobic bacteria operate

Wire mesh to keep the leaves from blocking the channels

Serrated troughs to spread liquid over bed

Graded clinker on which aerobic bacteria operate

Humus tank, otherwise a closed land drain system or surface furrows

Septic tanks *continued*

Anticipative problems	Considerations
	and each area used in turn at 8 to 12 week intervals. See also *Location of tanks* 1 *Note* on previous page. Closed layouts of drains with perforated or slotted pipes provide more even distribution of the effluent.
5 Precast fibre-glass units	Check with the manufacturer of these units **1** the correct grade to be used. These will vary with the invert levels of the inlets and the type and condition of the soil around it. **2** the installation requirements, for example on dry sites 100 mm of concrete and a 75 mm bed of sand together with a backfill around the unit with selected soil or sand may suffice, but in wet ground, hardcore, a polythene membrane and 150 to 225 mm of semi-dry, rapid hardening concrete base together with a similar thickness concrete surround up to invert level may be required. **3** the capacities available as these may vary from 2750 litres for 1 to 4 persons to 9100 for 16 to 20 people, the minimum depth and diameter being approx 2.27 and 1.85 m respectively. **4** the biological filtration plants available. On a number of sites where normal gravity dispersal of the treated effluent in the ground is not possible. Also pumping plant which is available to lift the liquid up and dispose of it on higher or remote areas on the site but away from the buildings may be required. *Note* The big advantage of these units is the speed with which they can be installed. All installations, precast and built *in situ* should preferably be fenced in to keep out animals and unauthorised persons.

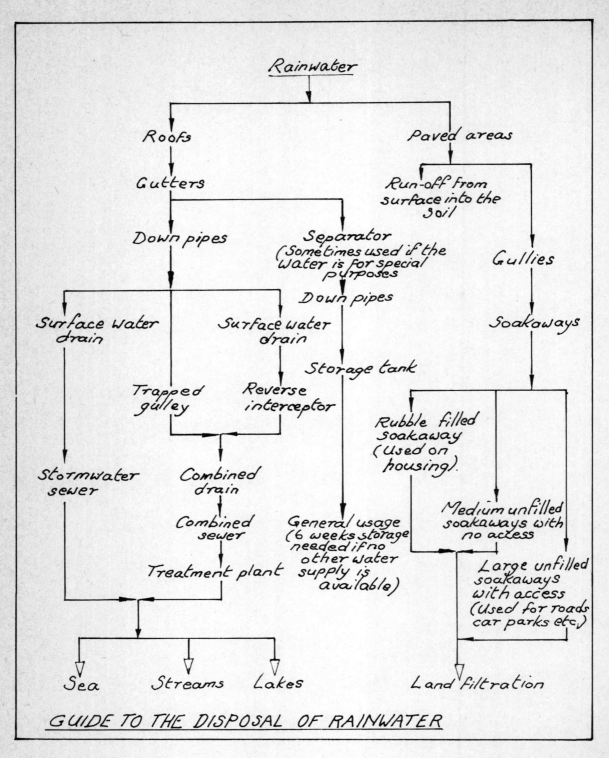

GUIDE TO THE DISPOSAL OF RAINWATER

6 Surface water disposal

All buildings must be provided with a means of collecting and disposing of rain water, this being a system of gutters, down pipes and drains. Down pipes may be external though it is often considered to be unsightly, or internal in which case all joints must be watertight and airtight otherwise problems arise as referred to later.

The sizing of gutters, down pipes and drains will depend upon the anticipated rainfall intensity which is usually taken as 75 mm hr, but roofs are constructed so as to form a kind of flat shallow bowl where overflows are advisable but for some reason cannot be tolerated then the figure of 150 mm/hr would be more reasonable.

The size of gutters and rainwater pipes will depend upon the flow load, gutter shape, depth of fall, shape of the gutter angles, the position of the outlets and the materials used. BRS Digests 188 and 189 give tables and examples on sizing, and some may be obtained from technical literature published by the manufacturers.

Gutters and down pipes

Anticipative problems	Considerations
	Gutters fixed after the completion of the roof
1 **Position of the eaves**	Ensure that the roof extends about halfway over the gutter. See figure 144.
2 **Falls in the gutter**	1 Ensure that gutters fall towards the outlet at a gradient of 1 in 300 or steeper. This will increase the flow capacity. See figure 143. 2 Ensure that the number of supports are adequate. Plastic gutters are lighter than other materials but need more supports to offset any tendency to sag. Also ensure that the gutters are set level from front to back otherwise the flow capacity will be reduced and spillage may occur.
	Gutters formed in situ with sheet materials
1 Maintenance	Ensure that gutters are at least 150 mm wide in order that a man may place his foot in them, or wider if it is anticipated that he may have to stand in them to carry out repairs to the roof.
2 **Internal leaks due to flooding**	Provide overflows for roofs enclosed on all sides to prevent a build up of trapped water and subsequent flooding resulting from an outlet which is possibly blocked with leaves and not visible without access. See figure 145. *Note* A larger outlet with a cage over will in some way reduce the problem, when there is no overlfow, but will not remove it.
3 **Roof stability**	Provide a trimmer for rafters running at right angles to the gutter.

123

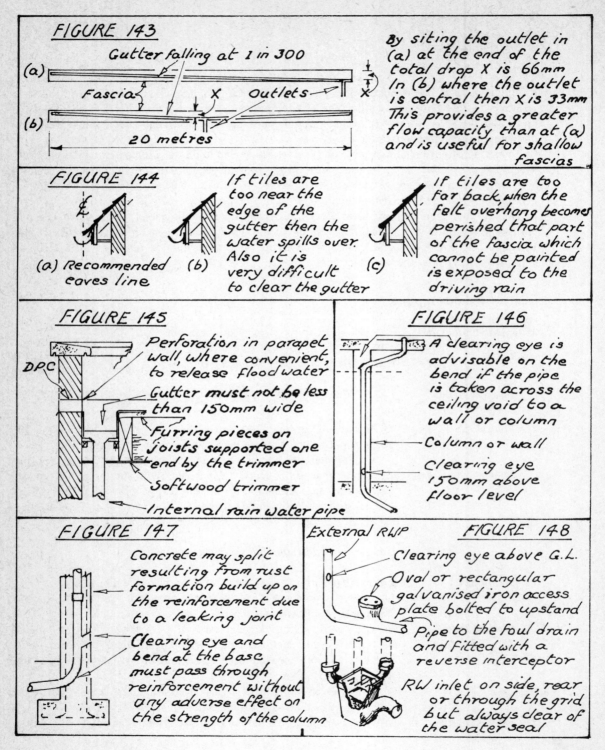

FIGURE 143

Gutter falling at 1 in 300

(a)

Fascia   X   Outlets

(b)

20 metres

By siting the outlet in (a) at the end of the total drop X is 66mm In (b) where the outlet is central then X is 33mm This provides a greater flow capacity than at (a) and is useful for shallow fascias

FIGURE 144

(a) Recommended eaves line   (b)   (c)

If tiles are too near the edge of the gutter then the water spills over. Also it is very difficult to clear the gutter

If tiles are too far back, when the felt overhang becomes perished that part of the fascia which cannot be painted is exposed to the driving rain

FIGURE 145

DPC

Perforation in parapet wall, where convenient, to release flood water

Gutter must not be less than 150mm wide

Furring pieces on joists supported one end by the trimmer

Softwood trimmer

Internal rain water pipe

FIGURE 146

A clearing eye is advisable on the bend if the pipe is taken across the ceiling void to a wall or column

Column or wall

Clearing eye 150mm above floor level

FIGURE 147

Concrete may split resulting from rust formation build up on the reinforcement due to a leaking joint

Clearing eye and bend at the base must pass through reinforcement without any adverse effect on the strength of the column

FIGURE 148

External RWP

Clearing eye above G.L.

Oval or rectangular galvanised iron access plate bolted to upstand

Pipe to the foul drain and fitted with a reverse interceptor

RW inlet on side, rear or through the grid but always clear of the water seal

Gutters and down pipes *continued*

Anticipative problems	Considerations

Notching of the joists to form the gutter should be avoided as it could affect the design depth and reduce the strength of these members. See figure 145.

Rain water pipes (Internal and external)

1 Blockages in the pipes

1 Ensure that the pipe diameters are never less than that of the gutter outlet.

2 Provide clearing eyes at bends and about 150 mm above ground or floor levels. See figure 146.

2 Leaking of the joints

1 Ensure that all internal joints are water and air tight as this pipe must not act as a vent to any drain especially in a combined system.

2 Ensure that all joints are watertight. Other leakages resulting from possibly blocked pipes may cause damp patches on the structure, stains and efflorescence, and increase any subsequent damage which may result from frost. See figure 147.

3 Damage from built in down pipes

Avoid built in down pipes. Metal pipes may eventually corrode and the joints leak, resulting in further rust formation on the concrete reinforcement together with its expansion causing the concrete to split. This could be accompanied with a staining of the internal finishes and damp patches on the faces of the walls or columns.

Note Long lengths of plastic pipe with solvent joints may avoid the above problem but

(a) the fixing of the pipe in space before the actual structure is formed around it is difficult, and, if it is passed down after the construction is complete, the joint between the drop and the outlet length going to the drain is now difficult to form and cannot be guaranteed to be watertight;

(b) problems may still arise when carrying the pipe through the structure where any displacement of the reinforcement could adversely affect the strength of the column. See figure 147.

Disposal of surface water

Rain water may be stored up for general use later on, or it may be disposed of through a system of drains to surface water sewers if available, combined sewers if approved by the Local Authority, or soakaways if site conditions permit it. Many areas have increased considerably faster than the sewage disposal installations which service them, so that generally soakaways are specified for the disposal of rainwater to avoid the likelyhood of overloading the sewage treatment works.

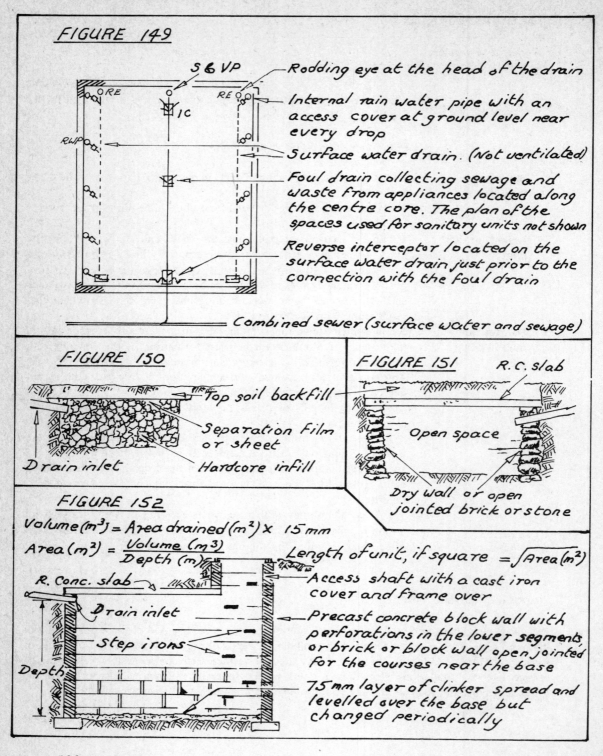

FIGURE 149

S & VP

ORE   RE

IC

RWP

Rodding eye at the head of the drain

Internal rain water pipe with an access cover at ground level near every drop

Surface water drain. (Not ventilated)

Foul drain collecting sewage and waste from appliances located along the centre core. The plan of the spaces used for sanitary units not shown

Reverse interceptor located on the surface water drain just prior to the connection with the foul drain

Combined sewer (surface water and sewage)

FIGURE 150

Top soil backfill

Separation film or sheet

Drain inlet

Hardcore infill

FIGURE 151

R.C. slab

Open space

Dry wall or open jointed brick or stone

FIGURE 152

Volume (m³) = Area drained (m²) × 15mm

Area (m²) = $\frac{Volume (m^3)}{Depth (m)}$

R. Conc. slab

Drain inlet

Step irons

Depth

Length of unit, if square = $\sqrt{Area (m^2)}$

Access shaft with a cast iron cover and frame over

Precast concrete block wall with perforations in the lower segments, or brick or block wall open jointed for the courses near the base

75mm layer of clinker spread and levelled over the base but changed periodically

126

Drains to surface water and combined sewers

Anticipative problems	Considerations
1 Blockages	Provide clearing eyes at the top end of surface water drains and near bends and junctions. Also rodding eyes are advisable at the base of each rain water drop and, where trapped gullies are to be used with long branch drains, in the gulley itself, if possible. See figures 135, 136 and 148.
2 Crossings with other drains	Check levels to ensure that, where it cannot be avoided, pipes can cross each other and yet still keep straight on plan and section. See figures 136 and 149.
3 Ventilation to be avoided	Provide a trapped gulley at the bottom of each rainwater pipe, or connect the drops with a surface water drain terminating with a reverse interceptor just prior to the junction with the foul and combined drains. See figures 136, 137 and 149. *Note* The rain water pipe must never act as a vent for foul drains because of the proximity of the gutters to openable windows.

Soakaways and drainage to and from them

A soakaway is a hole excavated for the collection of rain water from buildings before allowing it to seep away into the soil. It must be able to take the initial surge resulting from a sudden storm, an allowance of about one third of a normal hour's rainfall, after which the amount which enters the soakaway from the drains and the volume which seeps out from the base and lower sides will probably balance.

For domestic work see figure 150. This is usual and works satisfactorily. Because of the fill add at least 50% to the calculated capacity.

Larger, unfilled, sealed soakaways are illustrated in figure 151. For areas such as roads, carparks, etc, where the flow is more concentrated and flooding must be avoided see figure 152.

Anticipative problems	Considerations
1 Drainage requirements	1 Determine whether the soil will permit the water to drain away easily. Clay soils would not. 2 Check that the unit will be above the highest anticipated water table, otherwise the rainwater will not be able to drain away.
2 Location of the unit	1 Ensure that it is at least 5 m from any building otherwise it may contravene the Local Authority's regulations. Keep away from basements as it could put unnecessary pressure on the tanking. 2 Ensure that on sloping sites it is located on lower ground than the building, otherwise excavation costs would be unnecessarily costly.

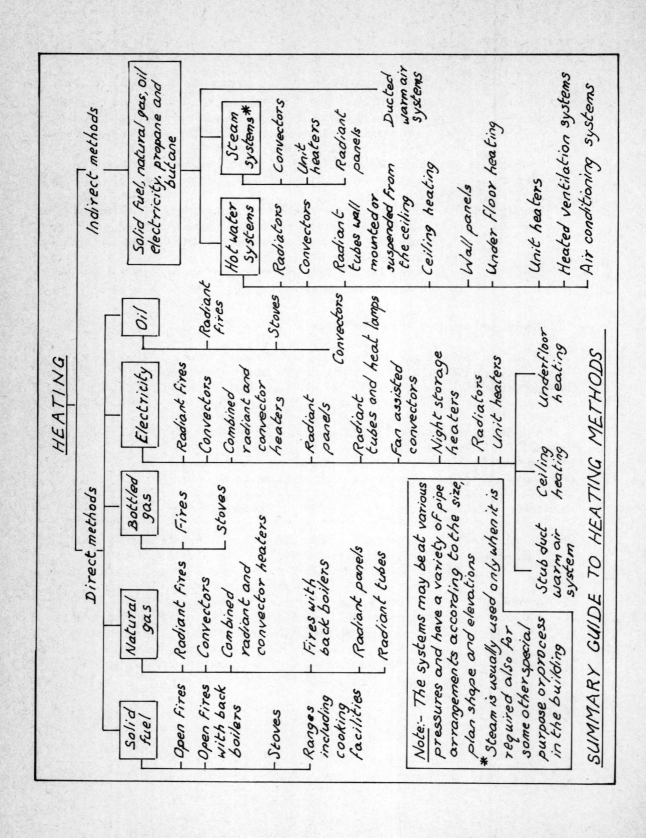

HEATING

Direct methods

Solid fuel
- Open Fires
- Open Fires with back boilers
- Stoves
- Ranges including cooking facilities

Natural gas
- Radiant Fires
- Convectors
- Combined radiant and convector heaters
- Fires with back boilers
- Radiant panels
- Radiant tubes

Bottled gas
- Fires
- Stoves

Electricity
- Radiant Fires
- Convectors
- Combined radiant and convector heaters
- Radiant panels
- Radiant tubes and heat lamps
- Fan assisted convectors
- Night storage heaters
- Radiators
- Unit heaters
- Ceiling heating
- Underfloor heating
- Stub duct warm air system

Oil
- Radiant Fires
- Stoves
- Convectors

Indirect methods

Solid fuel, natural gas, oil, electricity, propane and butane

Hot Water Systems
- Radiators
- Convectors
- Radiant tubes wall mounted or suspended from the ceiling
- Ceiling heating
- Wall panels
- Under floor heating
- Unit heaters
- Heated ventilation systems
- Air conditioning systems

Steam Systems ✱
- Convectors
- Unit heaters
- Radiant panels
- Ducted warm air systems

Note:- The systems may be at various pressures and have a variety of pipe arrangements according to the size, plan shape and elevations

✱ Steam is usually used only when it is required also for some other special purpose or process in the building

SUMMARY GUIDE TO HEATING METHODS

7 Heating

The heating of spaces to provide comfortable temperatures in which to work and live is achieved by one or more methods these being radiant or convection heating generally, and conduction in a few cases.

With radiant heating the object within the ray or beam is heated but the air between the object and the heat source is not. It is convenient where relatively instant warmth is required, for example, the result felt after switching on a normal electric bar fire when coming from a cold atmosphere into a cold room. It is suitable for intermittent use and in situations where convection would be uneconomic, for example, it would be a waste of energy to heat up the vast space within a high church when the warmth is only required at low level around the congregation and then possibly for just a few hours on specific days. Also it is advantageous in places such as factories where doors may be opening to the external atmosphere continuously, in that radiant heaters may be sited above the workers rather than try to heat the air which would spill out each time the doors were opened. Radiant heat is beamed heat, the direction being obtained by reflectors. See figure 153.

Convection heating is done by means of air circulation, the air near the heater being warmed, becoming less dense so lighter in weight, and being forced to rise by the cooler air away from the heater falling and displacing it. Heating up of a space takes less time using this method and the overall heating is better but it takes longer to appreciate and feel the heat available. The systems vary from the fan assisted air flows in ventilation and air-conditioning installations to the simple portable unit shown in figure 154.

Heating by conduction is usually reserved for cooking appliances, but some type of storage heaters use this method to build up a heat reserve during periods when demand is low and then dispel it gradually during the day. The electricity used to build up the store is offered at a cheap rate but unfortunately when most of the heat is required, that is possibly the evening, then most of the stored heat will have already been dissipated to the space around it. This system is probably the cheapest to install but is reputed to be costly to run. It is not advisable for areas subject to considerable solar heat gains as it can not be turned off once heated and the two heats could result in the space becoming too hot, the windows then being opened in order to cool it down, and the heat already charged for being wasted.

Heating installations may be classified as either direct or indirect. With direct heating the energy producing the heat is used directly in the space being warmed, the units being fixed or portable radiant or convector heaters, stoves etc or electric ceiling or floor warming systems. Although some direct heating appliances are provided with back boilers most have no provisions for heating water for sinks, etc, so separate units for this purpose are necessary.

Direct heating appliances and systems

Anticipative problems	Considerations
1 **Choice of unit** (a) Open fires	These provide a focal point desired by most occupants, and appreciated by the manufacturers of other fuel burning appliances as shown by their efforts to copy the log and coal effects. They are expensive because

Direct heating appliances and systems *continued*

Anticipative problems	Considerations

of the need for a chimney stack, chimney breast and a fuel store.

Open fires cause excessive ventilation rates resulting in high heat losses and inefficiency, and the need to refuel periodically and remove the ash daily is a disadvantage. The efficiency of the unit and the comfort levels may be improved quite considerably, as follows.

1 Locate the chimney breast on an internal wall. It will act as a heat store whilst the fire is burning and continue to dissipate warmth to the space around long after the fire is out. When more than one unit is involved they may be arranged in line or back to back. See figure 155.

2 Locate the stack terminal near the centre of the pitched roof and so reduce the need to thicken or supply stays for tall chimneys. See figure 155.

Note Shorter chimneys may be possible but if the terminal is in an area of variable pressure then problems of down draughts and smoke emmission into the room will arise. See figure 157. Some firms now produce a short balanced flue chimney stack, but it may not always match the finishes required.

3 Ensure that the chimney breast, hearth and the chimney stack dimensions and construction comply with Building Regulations part L some of which are illustrated in figures 155 anc 156.

4 Provide a fuel store large enough for at least three weeks supply and ensure that it is easily accessible for delivering coal and that there is enough room for a man with a sack on his shoulder, to turn without marking adjacent wall surfaces.

Note Smokeless fuel is lighter than other solid fuels and needs almost twice the space to accommodate the same weight.

5 Provide an adjustable metal throat restrictor in the entrance to the flue. This will reduce the rate of ventilation, help to prevent down draughts and reflect some of the heat which may otherwise have been lost in the flue, back into the room. See figure 156.

6 Provide an all-night burning unit. The hinged flap or spin wheel on the front gives a much more controlled rate of burning than the slotted plate and the ash can be reduced to a fine powder.

7 Provide air for combustion direct to the hearth from either the ventilated under floor space or the external air via airbricks and ducts from adjacent outside walls, and run below floor level.

Note If the outlet is near to the fire then the draughts, which are usually experienced by the replacement air being drawn across the room from cracks in doors and windows, will be reduced. If draughts are still apparant they could be the result of the warm air currents being cooled by contact with the cold window, dropping and returning across the floor to the fire, and not by cool exterior air being drawn in. See figure 158.

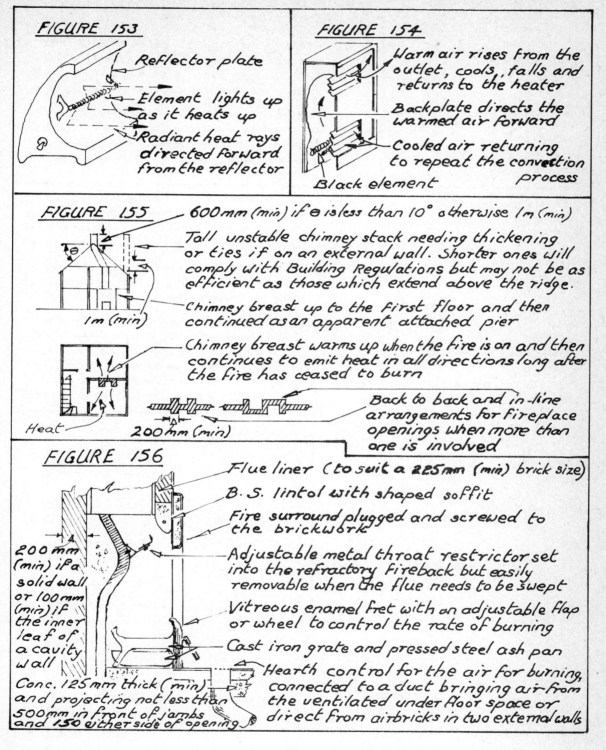

FIGURE 153

Reflector plate

Element lights up as it heats up

Radiant heat rays directed forward from the reflector

FIGURE 154

Warm air rises from the outlet, cools, falls and returns to the heater

Backplate directs the warmed air forward

Cooled air returning to repeat the convection process

Black element

FIGURE 155

600mm (min) if θ is less than 10° otherwise 1m (min)

Tall unstable chimney stack needing thickening or ties if on an external wall. Shorter ones will comply with Building Regulations but may not be as efficient as those which extend above the ridge.

Chimney breast up to the first floor and then continued as an apparent attached pier

1m (min)

Chimney breast warms up when the fire is on and then continues to emit heat in all directions long after the fire has ceased to burn

Heat

200mm (min)

Back to back and in-line arrangements for fireplace openings when more than one is involved

FIGURE 156

Flue liner (to suit a 225mm (min) brick size)

B.S. lintol with shaped soffit

Fire surround plugged and screwed to the brickwork

200mm (min) if a solid wall or 100mm (min) if the inner leaf of a cavity wall

Adjustable metal throat restrictor set into the refractory fireback but easily removable when the flue needs to be swept

Vitreous enamel fret with an adjustable flap or wheel to control the rate of burning

Cast iron grate and pressed steel ash pan

Hearth control for the air for burning, connected to a duct bringing air from the ventilated under floor space or direct from airbricks in two external walls

Conc. 125mm thick (min) and projecting not less than 500mm in front of jambs and 150 either side of opening

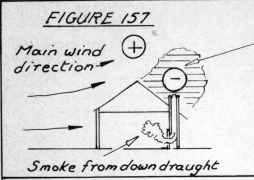

FIGURE 157

Main wind direction →

$\oplus$

Area of variable pressure. When the main wind blows air is induced from the shaded area giving it a negative pressure whilst the gust lasts

When the gust stops the air returns and flops back on to the stack and momentarily prevents smoke emission and may even cause it to blow back in to the room

Smoke from down draught

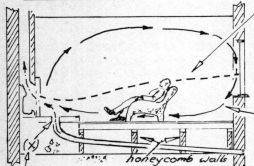

FIGURE 158

honeycomb walls

Fresh air from underfloor vented space or direct from outside

(X)

The dotted line indicates possible draughts due to combustion air being drawn in through cracks around the window or it could be a door

Fresh air drawn into the fire from X will bypass the room and eliminate the draughts mentioned above

Convection currents cooled by the cold glass and returning to the fire This draught is due to the cooling of the air when in contact with the glass and not air being drawn in. Double glazing will reduce or eliminate it

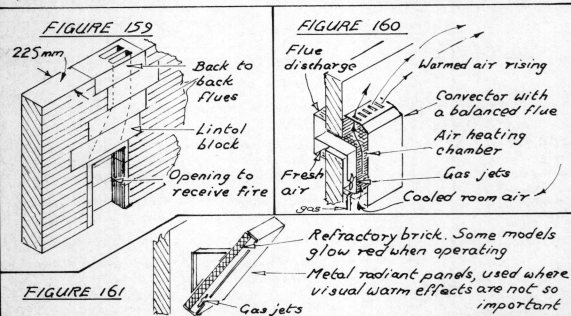

FIGURE 159

225mm

Back to back flues

Lintol block

Opening to receive fire

FIGURE 160

Flue discharge

Warmed air rising

Convector with a balanced flue

Air heating chamber

Gas jets

Fresh air

gas

Cooled room air

FIGURE 161

Gas jets

Refractory brick. Some models glow red when operating

Metal radiant panels, used where visual warm effects are not so important

Direct heating appliances and systems *continued*

Anticipative problems	Considerations
(b) Stoves	These may be inset or free standing, and will require a chimney and fuel store as before. If they are to discharge into an existing old flue then ensure that a chimney liner is provided otherwise the flue gases may attack the brickwork causing it to split or the stack to lean over and become unstable. Because stoves do not have the large opening as with open fires the acids in the flue gases are very concentrated.
(c) Gas radiant fires and convectors	These are mostly fixed units so must be provided with a flue which may be virtually built into the wall as shown in figure 159. Combined radiant and convection heaters are advisable for lounges thus making both types of heating available, but in bedrooms which may be un-attended for long periods then convection units, with the concealed burners, are best as there is less chance of any fires occuring resulting from bedclothes being left to near to the heater. See figure 160. *Note* Gas fires are a very useful second heat source for use when the central heating system has broken down or occassional warmth is needed.
(d) Gas radiant panels and tubes	Tubes and panels are used mainly at high level, for example in workshops where wall and floor space is restricted. Units available vary from single panels as shown in figure 161, suitable for toilet areas, to banks of heaters which glow red and are satisfactory for churches.
(e) Electric radiant and convector heaters	Electric heaters have advantages over others in that many are portable, they are very clean and do not need a flue or ventilation for cumbustion. The application is the same as for gas fires and convectors but as they may be obtained in various ratings, for example ½, 1, 1½, 2 and 3 kilowatts the fuse must be sized for the appliance otherwise the flex may overheat and start a fire. Fan assisted convectors can speed up the heating process but the fan noise may bother some users.
(f) Electric storage heaters	Storage heaters build up a reserve of heat at night when the demand for electricity is low and so obtain it at a reduced cost. They disperse it into the rooms during the day but unfortunately during the evenings when most heat is needed the reserves are at their lowest. They are clean but very heavy so should be sited near wall bearings and never mid span on timber joists. Radiant heat is always available whilst there is a reserve but fan assisted convection can also be obtained with some models. See figure 162. *Note* Because the emission is automatic once the heater core has been charged they are not advisable for large glazed areas facing south as the combined warmth from the unit plus the solar gain would cause overheating of the space. As neither can be turned off the windows would have to be opened and the heat built up overnight lost to the external air, though still to be paid for.
(g) Electric tubular heaters	These may be a single short tube for use in the airing cupboard, a line of them in the greenhouse or workshop, or a bank of them fixed at

Direct heating appliances and systems *continued*

Anticipative problems	Considerations

low level on the wall in offices. Air circulation around is necessary to prevent any overheating. See figure 163.

(h) Electric unit heaters

Unit heaters are fitted at high level and are useful in workshops, laboratories, etc, where overall heating is required but noise and any form of draught are not a problem.

(i) Other electric heaters

(j) Electric direct heating systems

1 Ceiling heating

Available also are radiant panels fitted below windows, and high level radiant heaters and heat light bulbs with sealed elements for bathrooms. Radiant panels are installed above the ceiling finishes. The on/off operation may be manual or by a time switch and the temperature controlled by a thermostat. It is clean, silent, invisible, needs no attention and can be automatic if so required. See figure 164. Consider also:

1 **the activities for which the space is used**   People who are standing a lot may complain that the temperature at head height is too high whilst others may complain of cold feet, especially if they are screened from radiation by a desk top;

2 **built-in furniture**   Whilst the floors and walls are unobstructed by pipes and emitters ensure that any furniture, built-in or otherwise, clears any ceiling panels by not less than 150 mm.

Note As the panels are not obvious a check is required before high cupboards are installed at a later date. If the panels are covered then the cables may overheat and a fire develop;

3 **running costs**   These may be high as the charges are based on the peak rates. Ensure that adequate insulation is provided above the panels to prevent the loss of heat upwards to areas other than the one being heated.

4 **Application**   It is suitable for houses, flats, offices and some shops and, if the pitch of the space is not too high, some schools and churches.

2 Underfloor heating

This is a form of heat storage in that the floor screed must absorb enough heat overnight to last through the following day. The cables may be run directly in the screed or through small ducts set in it, the latter making it easier to rewire at a later date if and when necessary. It is suitable for houses, shops, flats, offices, schools and churches, and may also be used as an anti-frost system for roads and garage drives. See figure 165. It is silent, clean, convenient and automatic but consider also:

1 **floor temperatures**   These must not exceed 24°C or this may cause people considerable discomfort to the feet. If at a later date very thick carpets are laid the heat emission upwards will be restricted and may be diverted downwards, and as the least heat is available when most needed in the evening it could be found to be inadequate. It provides a good even background heat but a direct heating appliance to supplement it if so required should be available;

FIGURE 162

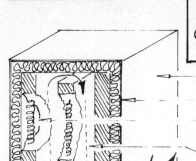

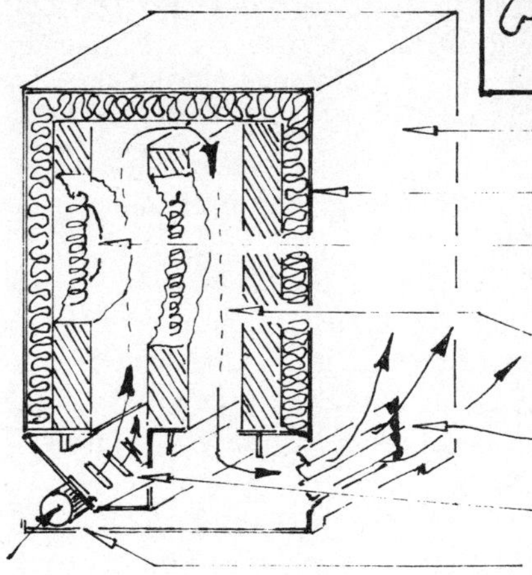

FIGURE 163

Spacer

Circular or oval tube heater containing the element and spacers. Tubes may be single or in banks

Metal cabinet and radiant heating surface

Insulation controlling the radiant heat

Heating element surrounded by refractory brick which causes the weight problem

Air gap for the warm air currents. Models without fans do not have this provided

Outlets for the convected warm air

Inlets for the cooled room air

2 speed electric fan. Fitted on some units

FIGURE 164

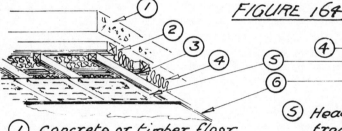

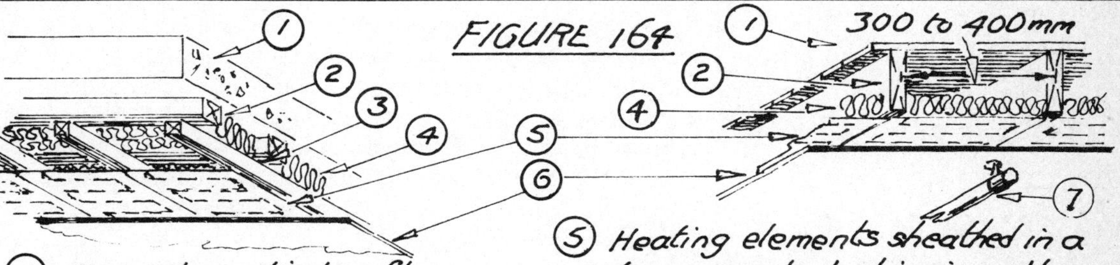

300 to 400mm

① Concrete or timber floor

② Timber joists or cross fillets fixed to the concrete soffit

③ Timber battens at 300 to 400 mm centres

④ Insulation - Thickness may vary dependant upon the heating requirements

⑤ Heating elements sheathed in a transparent plastic, in rolls 300 to 1200mm wide and 1.5 to 7.3m long, fixed to battens

⑥ Ceiling of 9mm plaster board with an 'Artex' Finish, 9mm timber or 6mm tempered hardboard

⑦ Light fittings fitted on the same point as the elements and at least 20mm clear of them

FIGURE 165

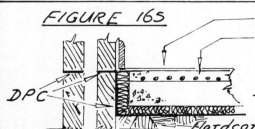

DPC

Floor Finish

Floor warming cables buried in the screed or in small ducts covered by it if provision for rewiring is required

Site concrete on a layer of insulation 25mm thick (minimum)

Hardcore blinded over with ash or lean concrete

Direct heating appliances and systems   *continued*

Anticipative problems	Considerations
	2 the differential movement of the concrete bays Ensure that each wiring panel is completely within a bay and does not fail due to possible tensions set up by movement of adjacent slabs;
	3 other circuits Ensure that any other cables which have to run in the screed are not adversely affected by the heat. Mineral insulated copper cable may be advisable;
	4 the insulation Ensure that the insulation will not or cannot be compressed so as to lose its heat insulation properties. Ensure also that the thickness is adequate to prevent undue heat loss to other areas. This will call for edge insulation to the slabs and possible thicker than normal layers below if it is known that very thick heavy carpets are to be used on the floor later;
	5 running costs The charges will be at the off peak rates as the heat is supplied during the night, but if it is to be used as background heating then peak charges will be made for any additional heat obtained by the direct heaters when used during the day and evening. This type of heating is not advisable for glazed areas facing south and subject to considerable solar gains for reasons already mentioned under *Electric storage heaters*, Page 133.
(k) Oil heaters	All oil heaters, if portable, should be of the self extinguishing type. Because of the large volumes of water vapour produced during the oil burning they are not recommended for situations where condensation might be a problem.
(l) Liquid gas	Provide space for spare gas containers for all bottled gas appliances and in some cases this store may need to be locked.

Indirect heating system

These systems usually comprise of a heat source which is a boiler fired by gas, oil, solid fuel and in a limited way, electricity, a distribution network of pipes carrying hot water and occassionally steam, or ducts carrying air, and a range of heat emitters which mostly are radiators, convectors or air grilles. They may be gravity operated, dependant on the difference in temperature of the flow (80°C) and the return (60°C) and their difference in density, or pump circulated.

Gravity operated hot water systems have advantages over the pumped systems in that they are silent, have no moving parts to wear out and are independent of other services. All pipes must be fixed to falls however, and they need to be of larger diameters than in pumped installations in order to overcome the friction set up by the pipe walls, bends, tees, etc. The difference in the temperature of the water in the flow and return will govern the difference in densities and control the rate of flow so that the lower the heat input is at source the slower the water circulates and the longer it takes to heat up a building.

The circulation in pumped systems is dependent on the electrical services and does give a much quicker heat up. Small diameter pipes can be used and the range of heat emitters for these systems is much larger. Where central heating is the only heat source it is advisable, in housing especially, to provide an alternative, for example a gas fire in the main area, to suse should the power fail or the pump cease to function.

Various pumped systems are available dependant on whether they are for domestic use or much larger establishments. The heat source may be in the building or located in an outhouse, but in both cases the problems of flues and fuel stores mentioned before must be considered.

Anticipative problems	Considerations
1 Choice of circulating system (a) Gravity	1 Ensure that all runs along the walls can be to falls. *Note* The minimum fall is 1 in 200. In long buildings the runs could be limited if the pipes are to be kept above the floor level. Also as the pipes are larger in diameter the space for falls and making connections is even more limited. 2 In areas where silence and independence of the other services is important this is ideal but the heat response can be very variable and often much slower than is required for comfort. See figure 166.
(b) Pumped circulation	This gives a much quicker response when areas are to be heated, falls are not necessary although they help to reduce pressures on the pump, and the smaller diameter pipes make it a much neater and possibly a cheaper installation. See figure 167.
2 Location of the pump	1 If it is on the flow pipe it must operate in the hottest water. It will provide a positive pressure and so possibly prevent any air entering into the system at high level through vents. 2 If located in the return then it will run cooler but could provide a negative pressure and air could be induced into the system at high level if the feed tanks are not sited high enough. 3 Ensure that the pump is fixed to the main structure and not to a partition wall as this may act as a sounding board and amplify any noise problems which may be present.
3 Choice of pipe arrangement or system	The system may be classified as one pipe or two pipe, and although the former may be a little cheaper the radiators become progressively cooler along the run and would need to be oversized in places in order to provide the necessary heat emmision. See figures 166 anc 168. Most systems tend to be two pipe installations where the hot flow water having passed through a radiator is returned direct to the boiler and not passed on to other radiators as in one pipe system.
(a) Small bore system	The pipes are 15 and 22 mm diameter copper usually, so are not difficult to conceal. The arrangement consists of a series of circuits and loops, is very popular for domestic use and easily adapted to automatic control. See figure 169.
(b) Mini- or micro-bore systems	This is usually for domestic installations. The pipes may be 8 or 10 mm diameter and as those running between the manifolds and the emitters are of soft copper they can be fitted into areas not possible with hard, rigid pipes. See figure 170.

137

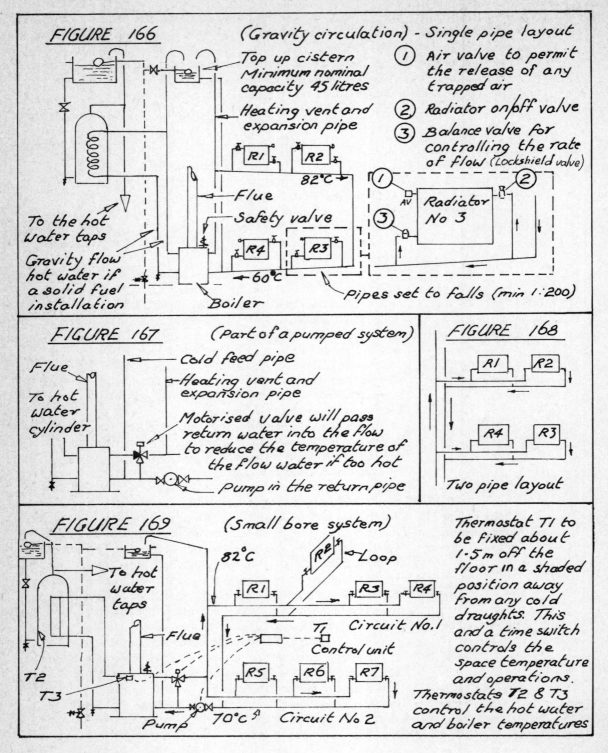

FIGURE 166   (Gravity circulation) - Single pipe layout

Top up cistern
Minimum nominal
capacity 45 litres

Heating vent and
expansion pipe

① Air valve to permit
the release of any
trapped air

② Radiator on/off valve

③ Balance valve for
controlling the rate
of flow (Lockshield valve)

R1   R2
82°C

① AV   Radiator No 3   ②

③

Flue
Safety valve

R4   R3
← 60°C

Boiler

To the hot
water taps

Gravity flow
hot water if
a solid fuel
installation

Pipes set to falls (min 1:200)

FIGURE 167   (Part of a pumped system)

Flue

To hot
water
cylinder

Cold feed pipe

Heating vent and
expansion pipe

Motorised valve will pass
return water into the flow
to reduce the temperature of
the flow water if too hot

Pump in the return pipe

FIGURE 168

R1   R2

R4   R3

Two pipe layout

FIGURE 169   (Small bore system)

82°C

R2   Loop

To hot
water
taps

R1   R3   R4

Flue

T1
Control unit   Circuit No.1

T2

T3

R5   R6   R7

Pump   70°C   Circuit No 2

Thermostat T1 to
be fixed about
1·5 m off the
floor in a shaded
position away
from any cold
draughts. This
and a time switch
controls the
space temperature
and operations.
Thermostats T2 & T3
control the hot water
and boiler temperatures

138

Indirect heating system *continued*

Anticipative problems	Considerations
	Note As the pipes are smaller than with other systems the installation is not necessarily any cheaper as two pipes are required from the manifold to each radiator and these may be longer than would be otherwise used. Also the chance of blockages must be higher when one considers the small diameters used and the amount of black sludge which can be removed when radiators are washed out. Some of this sludge may be reduced in volume or prevented from forming by the use of special inhibitors in the water when filling up. Great care is needed when installing soft pipes to ensure that the bores are not distorted by flattening, and the system needs to be flushed out several times on completion to ensure that no residue is left in the pipes. They are very easy to conceal but care is needed to ensure that they are not damaged when covered up or at a later date because their location is not known.
(c) Upfeed system	This system is useful on buildings where the various sections of the structure rise to different heights. The feed cistern must be above the highest emitter and all high spots must be vented. See figure 171.
(d) Drop system	This is useful for regular shaped buildings with roofs all on one level, where long runs can be made at high and low level the two being joined by drops run in vertical ducts, or internal angles adjacent to columns. See figure 172.
(e) Ring system	Long single storey buildings where the runs are not impeded by internal columns are suitable for this system. Where columns are essential then it would be better for this system if they could be sited externally in order to maintain straight pipe runs.
(f) Ladder system	This is an extension of the last for use in the higher buildings where each floor has its own ring, and drops, other than those in specially designed service cores, are not practical or even possible. See figure 173.
4 Boiler rating	This will be calculated in Btus (British thermal units) or W (watts). It will be approximately the summation of the heat losses from the building plus the amount of heat used in providing hot water plus a percentage, between 15 and 30 generally, to cover the inefficiency of the boiler itself. *Note* Btu/hr is 0.293 W. 1W is 3.412 Btu/hr.
5 The position of the feed tanks	Ensure that feed tanks are always above the level of any horizontal heating runs and always above the highest emitter. If they are not and it is an open system then some would not fill up.
6 Expansion of the water when hot	1 When it is an open system ensure that the heating vent and expansion pipe extends above the top of the feed cistern for not less than one twenty fourth of the head measured from the centre of the boiler, and that any discharge from this pipe runs back into the feed cistern. See figure 170. 2 If sealed systems are to be used then ensure that the correct

139

Indirect heating system   *continued*

Anticipative problems	Considerations

expansion vessel is provided. See figure 174.

Note   Any expansion of the water is taken up by the pocket of nitrogen in the vessel being compressed. This will also increase the boiling temperature of the water. Consult the manufacturer on this otherwise if the wrong unit is selected the radiators could become too hot and burn people who touch them. Also because of the extra heat the heat exchanger in the hot water tank may have to be reduced in size.

7   Expansion of the pipes

1   Provide some change of direction on the long runs, and space around the pipe at the change point in order that it can move as the temperature varies. See figure 175.

2   Where pipes are buried in floors then provide channels in the screed and insulation around the pipes, making allowances for any crossovers. Copper pipes buried in the screed should be covered with high temperature PVC. See figures 176.

3   Where pipes are to be buried in plaster then ensure that they are screened off with a strip of expanded metal lath before the plaster is applied, especially if the cover is less than 22 mm thick.

Note   Burying pipes in plaster should generally be avoided as it is always difficult to trace them and they can accidentally be damaged by the occupier plugging and fixing to the wall for other reasons resulting in an unnecessarily large and expensive repair job, depending on the kind of finishes existing.

8   Obstruction by columns

Where horizontal wall runs are anticipated then any proposed columns on this run are best located externally. If pipes have to be taken around the columns then the resistance to flow is increased considerably. Also it does not present a very neat or pleasing appearance. Where columns must be internal then one of the following may be adopted.

1   Provide cardboard tubes or polystyrene blocks in formwork where holes may be anticipated, before placing the concrete, ensuring that they are well anchored and will not float. These may easily be removed when the pipes are fitted. See figure 177(a).

2   Run the main flow and return in the floor and top off various emitters. See figures 177(b) and (c).

Note   The screed must be thick enough to allow the pipes to cross over where required, and if a normal joisted floor is used then the pipe run and the direction of the joist must be the same.

3   Run pipes in corridors above a suspended ceiling and tap off to the units. This may be a little more costly because of the runs across the ceiling but can be justified more easily than can a false ceiling in a complete room to perhaps hide a couple of pipes. See figure 177(d).

9   Ventilation for combustion

All gas, oil and solid fuel boilers must have permanent ventilation and flues complying with Building Regulations parts L and M. Regarding

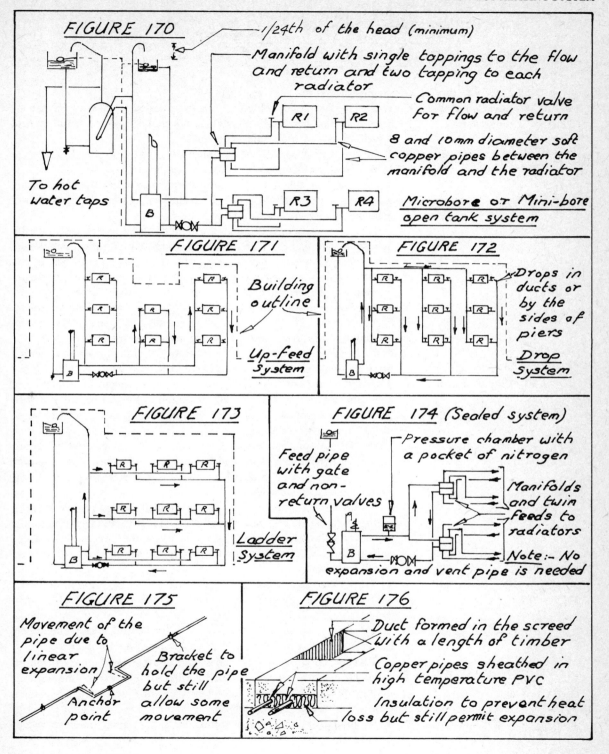

FIGURE 170

1/24th of the head (minimum)

Manifold with single tappings to the flow and return and two tapping to each radiator

Common radiator valve for flow and return

8 and 10mm diameter soft copper pipes between the manifold and the radiator

Microbore or Mini-bore open tank system

To hot water taps

R1 R2

R3 R4

B

FIGURE 171

Building outline

Up-feed System

B

FIGURE 172

Drops in ducts or by the sides of piers

Drop System

B

FIGURE 173

Ladder System

B

FIGURE 174 (Sealed system)

Pressure chamber with a pocket of nitrogen

Feed pipe with gate and non-return valves

Manifolds and twin feeds to radiators

Note:- No expansion and vent pipe is needed

B

FIGURE 175

Movement of the pipe due to linear expansion

Bracket to hold the pipe but still allow some movement

Anchor point

FIGURE 176

Duct formed in the screed with a length of timber

Copper pipes sheathed in high temperature PVC

Insulation to prevent heat loss but still permit expansion

141

Indirect heating system *continued*

Anticipative problems	Considerations

the position of the vents see Figure 50.

10 Location of the boiler

1 Ensure that models which are not room sealed are near to permanent flues and the hot water cylinder, and that space for maintenance, refuelling, etc, as applicable, is adequate.

2 Ensure that noise will not be a problem.

Note Noises which may bother some people are:

(a) wind in the flue. This is probably worse on bungalows using wall hung traditionally flued boilers. Bird songs which are not always pleasant are also easily carried down short flues;

(b) the on/off operation of the boiler caused by the solenoid, the lighting of the flame and in some cases the actual burning. Back boilers are probably quietest but are generally in rooms where noise is not wanted. Wall flame oil burners may be noisy due to both the combustion and the electric motor spinning the oil delivery pipe.

Note Where noise cannot be tolerated then the solid fuel continually burning boiler could be used, but should the daily maintenance, refuelling, etc, for this also be introlerable, then it is better to site the boiler in an outhouse or else separate from the inhabitable rooms with a brick or block wall not less that 225 mm thick.

11 Choice of heat emitter
(a) Exposed pipes

The type chosen will depend upon where it is located, its specific use and appearance. These are usually a section of the normal run but with larger diameter pipes to provide a greater heating surface. These tubes may be located:

(a) around and below roof lights and clerestory windows to reduce possible down draughts and condensation which may occur at high level;

(b) in airing cupboards where there is no hot water cylinder, it having been located elsewhere;

(c) in store rooms and workshops at high level where convection methods are costly and impractical. A metal plate with insulation on the top diverts the heat down to where it is wanted. An alternative is a radiant panel where the pipes are concealed. See figures 178 and 181 and *Note* 11(g).

(b) Radiators

1 **Panel radiators** These are steel units in common use in domestic and commercial premises and may be of the single or double panel types.

Note Double panel radiators have twice the heating area of single ones but emit less than twice the heat per m^2 for the same wall area.

2 **Thermal radiators** Some units have convection channels fitted on the back or between panels which increases the emission rate without taking up extra wall space. See figure 179. In time it might become a dust trap if not cleaned periodically.

3 **Hospital radiators** The unit must be smooth and have clean sweeps

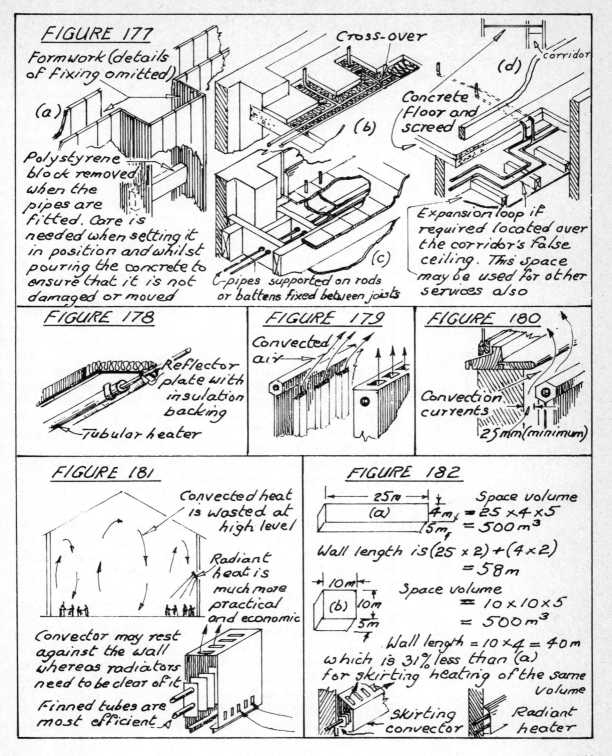

FIGURE 177

Formwork (details of fixing omitted)

(a)

Polystyrene block removed when the pipes are fitted. Care is needed when setting it in position and whilst pouring the concrete to ensure that it is not damaged or moved

Cross-over

(b)

(d) corridor

Concrete floor and screed

Expansion loop if required located over the corridor's false ceiling. This space may be used for other services also

(c)

C-pipes supported on rods or battens fixed between joists

FIGURE 178

Reflector plate with insulation backing

Tubular heater

FIGURE 179

Convected air

FIGURE 180

Convection currents

25 mm (minimum)

FIGURE 181

Convected heat is wasted at high level

Radiant heat is much more practical and economic

Convector may rest against the wall whereas radiators need to be clear of it

Finned tubes are most efficient

FIGURE 182

25m

(a) 4m 5m

Space volume
$= 25 \times 4 \times 5$
$= 500 \, m^3$

Wall length is $(25 \times 2) + (4 \times 2)$
$= 58 \, m$

10m 10m
(b) 5m

Space volume
$= 10 \times 10 \times 5$
$= 500 \, m^3$

Wall length $= 10 \times 4 = 40 \, m$
which is 31% less than (a) for skirting heating of the same volume

skirting convector

Radiant heater

Indirect heating system *continued*

Anticipative problems	Considerations

between surfaces rather than angles to ensure that cleaning is easy and no cracks are provided for harbouring germs.

4 Column types These are usually of cast iron and used on larger buildings where a long life is of more importance than appearance.

Note

1   Fit radiators at least 25 mm from wall face to provide both radiant and convected heat.

2   Locate radiators below windows as it helps to reduce draughts caused by convection currents being cooled by the cold glass. Also it reduces wall staining and condensation, but has the disadvantage that long curtains cannot be used without shutting off the heating surfaces from the room. See figure 180.

3   Radiator shelves when fitted do not prevent staining but cause it to occur at a higher level.

4   Avoid the use of metalic paints on radiators as this reduces the heat emission.

(c) Wall convectors

These generally are smarter and neater in appearance than radiators and may be fitted right back against the wall. Some have finned tubes which provide a better rate of heat emmission than the plain ones. Convectors give a quicker overall heat up but are uneconomic for large high spaces such as churches, garages, etc. See figure 181.

(d) Skirting heaters

These may replace the normal skirting but negotiating columns can be difficult and expensive. They may be radiant or convector types and form a warm air blanket around the room.

Note The building shape can considerably affect the length of heater required, for example, two buildings may have the same floor areas and be of the same volume yet one will require over 40% more run to achieve the same result. See figure 182.

(e) Ceiling panesl

Some are exposed suspended radiant panels used at high level in workshops the temperatures varying with the mounting height, whilst others are hairpin shaped coils fitted below the structural slab in a false ceiling. See figure 183(a). Some people may find this oppressive as heating from above destroys the normal convection currents and any feeling of freshness which they provide, but it does leave floors and walls clear. Also dust layers may be more obvious.

(f) Floor panels

Hairpin pipe arrangements are set in the floor screed so the heating up period is slow as the mass must heat up first. The temperature must be restricted to 24°C otherwise it can cause discomfort. It is good for churches if open each day, homes for the aged and disabled, and schools where children spend much time playing on the floor. It should not be used in areas subject to solar gains because the two together could cause over-heating, discomfort and the subsequent opening of windows and

FIGURE 183

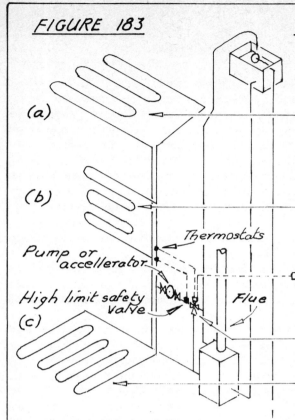

(a)

(b)

Thermostats

Pump or accellerator

High limit safety valve

(c)

Flue

Panel heating

Pipe coils of 12mm bore in hair pin or sometimes grid systems of panel with not more than 50m of pipe per panel

<u>Ceiling panels</u> – Pipes at 150mm centres bedded in the soffit of the concrete floor slab or suspended from it on hangers

<u>Wall panels</u> – Pipes at 150 mm centres are welded on steel flats bolted into the brickwork and covered with expanding metal lath before plastering to avoid the plaster cracking when heating

External thermostat located in a sheltered and permanently shaded position on the wall

Mixing valve controlled by the external and internal thermostats

<u>Floor panels</u> – Pipes at 150 and 225mm centres are buried in the screed on the concrete floor

<u>Note</u>:– Where large areas are involved then each system or repeat of layout may be controlled individually and have it's own pump or accellerator sited just after the flow header near the boiler

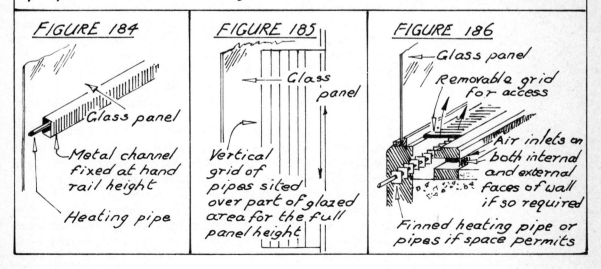

FIGURE 184

Glass panel

Metal channel fixed at hand rail height

Heating pipe

FIGURE 185

Glass panel

Vertical grid of pipes sited over part of glazed area for the full panel height

FIGURE 186

Glass panel

Removable grid for access

Air inlets on both internal and external faces of wall if so required

Finned heating pipe or pipes if space permits

Indirect heating system *continued*

Anticipative problems	Considerations

wastage of heat. See figure 183(b).

(g) Wall panels

Exposed wall panels may be used in workshops, garages, etc, and low temperature ones may be used in halls and on landings. Concealed ones are usually sited on stairways and landings where the heat emission will not be impeded by furniture. Temperatures should not exceed 29°C otherwise it may cause the plaster to crack. See figure 183(b).

(h) Unit heaters

These are located at high level in workshop areas or others where fan noise and the currents of air caused by the moving blades is on no consequence. Oscillating models are available for the larger areas where the number used is limited.

(i) Heating of spaces where areas of glass are involved

1 Long radiant panels fixed to the glass and acting as arm rests for viewers. See figure 184.

2 Banks of vertical tubes over the glazed areas. See figure 185.

3 Low level or below floor level type of convector heater. This has an advantage sometimes in that it may also include ventilation. through the outer leaf of the cavity wall. See figure 186.

12 **Temperature control location**

Temperature is controlled by thermostats which may be a central unit located in a general area such as the hall or corridor, or individual units sited in particular zones, rooms or on the actual emitters themselves such as thermostatic valves. In larger establishments the main thermostat may be located on an external face of the building.

1 Ensure that all air temperature sensing thermostats, external and internal, are never in the sun or subject to excessive cold draughts otherwise the warmth from the sun or the cold air blast will cause the heating system to shut off or open up before conditions internally are satisfied. *Note* If the unit is sited internally on a wall near a south facing window or glazed door the sun rays coming through the glass on to it may supply enough warmth to cause it to shut the system down even though the rest of the house is still cold, and it could remain off all the time the sun is on it. If at other times the window is open in order to provide some ventilation then the cool draught may cause the system to start up even if the other rooms in the building are hot and it may stay on if the window is not closed thus causing excessive heating and discomfort in some areas.

2 Ensure that it is located in a common area such as a hall or corridor. If it is controlling the temperature of several rooms and is located in one occupied by a fresh air enthusiast who insists on windows being open all day then again it could cause the heating to stay on longer than needed and cause overheating in the other rooms.

3 Ensure that it is not near to a radiator as this could provide a false signal and it will fail to call for heat assuming that the required temperature has been reached overall, even though the other rooms in the

Energy source	Appliance or system considered	Direct heat	Indirect heat	Fuel delivery	Fuel storage	Ash removal	Flues & chimney	Ventilation	Gas meters	Electricity	Cold water feed cisterns	Hot water cylinders	Water pipes	Water pump	Air ducts	Fan
Electricity	Warm air systems	I	✓×	I	I	I	I	I	I	✓	I	✓—	—	✓—	✓—	✓	✓
Electricity	Ceiling heating	□×	✓	I	I	I	I	I	✓	I	I	I	I	I	I	I	
Electricity	Underfloor heating	□×	✓	I	I	I	I	I	✓	I	I	I	I	I	I	I	
Electricity	Night storage heaters	□×	✓	I	I	I	I	I	✓	I	I	I	I	I	I	I	
Electricity	Fires and convectors	□×	✓	I	I	I	I	I	✓	I	I	I	I	I	I	I	
Electricity	Tubes & lamps	✓×	I	I	I	I	I	I	I	I	I	I	I	I	I	I	
Oil	Warm air systems	I	✓	⊕	⊕	✓	✓	✓‡	I	✓	✓—	✓—	✓—	✓—	✓		
Oil	Boiler	I	✓	⊕	⊕	✓	✓‡	I	✓	✓	✓	✓+	I				
Oil	Stove	✓×	I	✓	✓	✓	✓	I	I	I	I						
Butane	Boiler	I	✓	◇	◇	✓	✓‡	✓	✓	✓	✓+	I					
Butane	Stove	✓×	I	✓	✓	✓	I	I	I	I	!						
Gas	Warm air system	I	✓	I	I	✓	✓‡	✓#	✓	✓—	✓—	✓—	✓	✓			
Gas	Boiler	I	✓	I	I	✓	✓‡	✓#	✓	✓	✓+	I					
Gas	Radiant panels	✓×	I	I	I	✓	✓‡	✓#	I	I	I						
Gas	Fires with back boilers	✓	I	I	I	✓‡	✓#	✓	✓	✓+	I	I					
Gas	Fires and convectors	✓×	I	I	I	✓	‡	✓#	I	I	I						
Solid fuel	Boiler systems	I	✓	✓	◇✓	✓	✓‡	✓+	✓	✓	✓○+	I					
Solid fuel	Stoves	✓×	I	✓	◇✓	✓	✓‡	I	I	I	I						
Solid fuel	Fires with back boilers	✓	I	✓	◇	✓	‡	+	✓	✓	✓	✓○+	I	I			
Solid fuel	Open fires	✓×	I	✓	◇✓	✓	✓	✓‡	I	I	I	I	I				

Legend:
- ✓ – Items to be considered with the appliance or system.
- + – Electric pump required if gravity circulation is not used.
- ‡ – Ventilation is required for combustion.
- ‡# – Ignition may be from mains or battery operated.
- ◇ – Bituminous coal requires less space than smokeless fuel.
- ‡ – Minimum storage recommended is 3 weeks (domestic), otherwise 6 weeks. Lorry access is required for large buildings.
- ○ – Hard standing required off the public highway.
- ◇ – Avoid using catchpits as leaking gas may be trapped.
- ✓ – Hot water supply may not be incorporated.
- ⊕ – Supplied by tanker hose. Remote tanks need a permanent site pipe line installed.
- ⊕ – 3 weeks storage advised, above ground and over a catch pit.
- □ – Fan assisted models are available.
- ‡ – Large buildings may need a transformer externally or internally with a CO₂ room.
- × – Domestic hot water is supplied separately.
- ○ – Heating of domestic hot water must be by gravity circulation.

FIGURE 187. General overall appraisal

Indirect heating system   *continued*

Anticipative problems	Considerations
	building are still cold.
	4 Ensure that areas in buildings subject to uncontrolled heat gains, for example large glazed areas facing south and west, or kitchens where the cooker and electrical equipment is used considerably, that any emitters which are installed are fitted with thermostatic radiator valves which can stop the emission from that unit when that and the uncontrolled heat gain combined reach the required room temperature, and so prevent overheating in that particular area.
13 Other considerations	An overall appraisal of the types of heating unit and systems, together with the necessary ancillary requirements for their installation and operation is shown in figure 187.

Warm air heating

Warm air heating is suitable for all open plan types of building providing the pitch of the rooms or spaces is not too high, for example whilst it is quite acceptable for some types of housing and offices it would not be economic if used in lofty warehouses unless initially the air was blown down into the gangways along which people walk or work.

It differs from water systems in that the ducts form an integral part of the building and must be considered at the planning stage. It cannot be added when the structure is complete as can most water systems.

Warm air may be supplied direct through individual heaters which are sited in the areas being heated but this is usually adopted by industrial establishments. See figure 188. Indirect systems may have the heater located in a ventilated cupboard adjacent to the rooms requiring heat and connected to them by stub ducts, as used in housing and shown in figure 189, or the heater may be remote and sited in a plant room, the warm air being conducted around the building by ducts as used in offices, stores and industrial buildings. See figure 190.

Generally direct heaters use gas or electricity to warm the air, but for indirect systems solid fuel and oil are satisfactory these fuels being used to heat water initially which is passed to heating coils sited in the ducts which in turn warm the air as it passes through. See figure 191.

Warm air systems generally provide a more even temperature distribution overall and create a positive pressure within the rooms thus reducing draughts. They can be used as a general background heat source with radiant heat units to top up selected areas if required. In summer they can circulate air with the heating off, thus providing the effect of cooling the rooms.

Anticipative problems	Considerations
1 Choice of flow pattern for the heater	(a) *Downward flow* — This is suitable for installation in cupboards where the floor space is otherwise limited. See figure 192(a).
	(b) *Upward flow* — This is satisfactory for installations in basements. See figure 192(b).

148

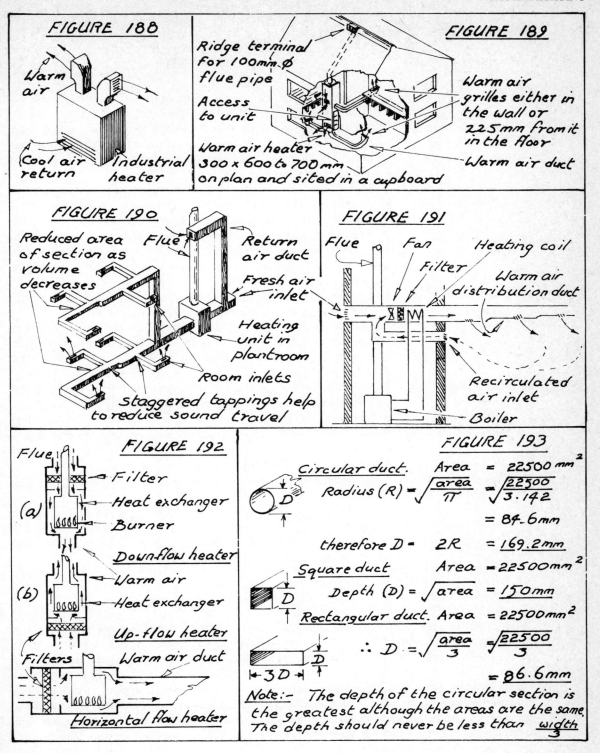

FIGURE 188

Warm air

Cool air return

Industrial heater

FIGURE 189

Ridge terminal for 100mm ∅ flue pipe

Access to unit

Warm air heater 300 × 600 to 700 mm on plan and sited in a cupboard

Warm air grilles either in the wall or 225mm from it in the floor

Warm air duct

FIGURE 190

Reduced area of section as volume decreases

Flue

Return air duct

Fresh air inlet

Heating unit in plantroom

Room inlets

staggered tappings help to reduce sound travel

FIGURE 191

Flue

Fan

Filter

Heating coil

Warm air distribution duct

Recirculated air inlet

Boiler

FIGURE 192

Flue

(a)
Filter
Heat exchanger
Burner

Down-flow heater

(b)
Warm air
Heat exchanger

Up-flow heater

Filters

Warm air duct

Horizontal flow heater

FIGURE 193

Circular duct.   Area = 22500 mm²

$$\text{Radius } (R) = \sqrt{\frac{area}{\pi}} = \sqrt{\frac{22500}{3.142}}$$

$$= 84.6 mm$$

therefore D = 2R = 169.2mm

Square duct   Area = 22500mm²

$$\text{Depth } (D) = \sqrt{area} = 150mm$$

Rectangular duct. Area = 22500mm²

$$\therefore D = \sqrt{\frac{area}{3}} = \sqrt{\frac{22500}{3}}$$

$$= 86.6mm$$

Note:- The depth of the circular section is the greatest although the areas are the same. The depth should never be less than $\frac{width}{3}$

Warm air heating *continued*

Anticipative problems	Considerations
	(c) *Horizontal flow* – This is designed for use above cupboards and below working tops. See figure 192(c).
2 Ventilation and removal of the combustion gases	Ensure that adequate ventilation is provided for gas direct units located in cupboards and that the flue, if not sited on an external wall, can continue satisfactorily up through any rooms or walkways which may be immediately above.
3 Location of ducts	1 Ensure that the joists run in the same direction as that required by the ducts at that level, or where this is not possible provide trimmers to support the joist ends as in figure 19(a). 2 When ducts run below floor levels ensure that the voids above the false ceiling are deep enough to accommodate them. Runs sited along corridors might be more satisfactory and less expensive depending on the position of the room inlets.
4 Condensation in the ducts	Ensure that ducts which run in ventilated spaces are lagged to prevent condensation on them and unnecessary heat losses on the run.
5 Choice of duct section	Ensure that ducts, other than stub ducts, are if possible square or rectangular in section rather than circular. These do not have to be as deep to permit the same volume of air flow but the ratio of width to height should be from 1 to 1 to 3 to 1. See figure 193.
6 Maintenance provisions	Ensure that there is adequate space around the unit to service it, and remove and replace filters periodically as they become clogged with dirt.
7 Domestic hot water	Some models incorporate provisions for domestic hot water, but where they do not then ensure that additional plant to cover this need is provided.
8 Other essential considerations	(a) Duct sizes. Consult the proposed suppliers. (b) Flue sizes for heaters other than electric, and compliance with Building Regulations. (c) Fuel delivery and storage for other than gas and electric heaters. (d) Joist runs, joist depths, weight, provision for access and the location of room inlets and outlets as these cannot be changed as can a radiator once the installation is complete.

Energy conservation

Heating uses up our natural resources which are rapidly running out so the conservation of energy should always be considered when any heating installation is proposed. As heat is emitted into a space so it leaks away through the walls and roof and constantly needs replacing. Insulation can restrict the rate of heat loss so should always be considered in conjunction with heating installations whether direct or otherwise and it may be in different forms or applied in various ways.

Insulation of roof space   This is easy and relatively cheap but care is needed where pitched roofs are concerned to retain some ventilation and reduce the possible passage of water vapour from bathrooms and kitchens through the perforations in the ceilings made for pipe runs. Water vapour escaping into the roof space will increase the humidity in that area and may cause the roof timbers to start rotting, and condensation to occur in the conduits over these rooms which can lead to electrical failures. Whilst insulation is to be encouraged care is needed as the better the insulation the greater the problem just mentioned can become.

Wall insulation   Existing cavity walls may be filled, but as this destroys the initial purpose of the cavity, guarantees are necessary to ensure that once it has been filled rain penetration cannot occur, or should it develop later that the problem can and will be put right by the contractor or any other contractor taking over his responsibilities.

Note: Today many existing timber windows are being replaced by uPVC and aluminium ones. The new windows are the same width as the opening in order to avoid cutting away the brickwork, and so are not recessed into the cavity as fitted originally. Should the cavity have been filled prior to the installation of the new windows it is essential to ensure that the window contractors fill the voids left between the back of the new units and what was the back of the old windows otherwise rainwater seeping through the outer leaf can be trapped in these pockets. The voids on either side of each window may measure 75 x 50 mm or more on plan, and extend for the full height of the frame. Once in the cavity the water will seep through the inner skin causing damp patches on the wall internally. Whilst the fault may appear to be the responsibility of the cavity fill contractors, on investigation it may be found that any guarantee they may have given is now valueless because the wall is not as they originally left it.

With solid walls insulation can be applied internally or externally but the latter is very expensive as it has to be protected from the weather and against damage by impact. Whilst it may enhance the appearance of older buildings the cost is possibly only justified in the case of buildings heated with electricity or those which need to be heated continuously throughout the day and night. On new work an effort should be made to retain the cavity and provide the insulation by the use of materials with low heat transfer properties for the inner skin.

Double glazing   This does not conserve as much heat as is first anticipated but the reduction in draughts, condensation and nuisance from external noise is most noticeable when double glazing is installed. The costs may not be justified by the savings in heating bills but the other benefits just mentioned may be considered a just reward for the expense involved for the installation.

Building shapes   Avoid the use of re-entrant walls on plan as this reduces the area of the enclosed space but increases the run of wall around it, and as the bulk of the heat loss can be through the wall surfaces they need to be kept to a minimum. Basically square buildings are best for conserving heat but their external appearance is obviously lacking in any visual interest or originality.

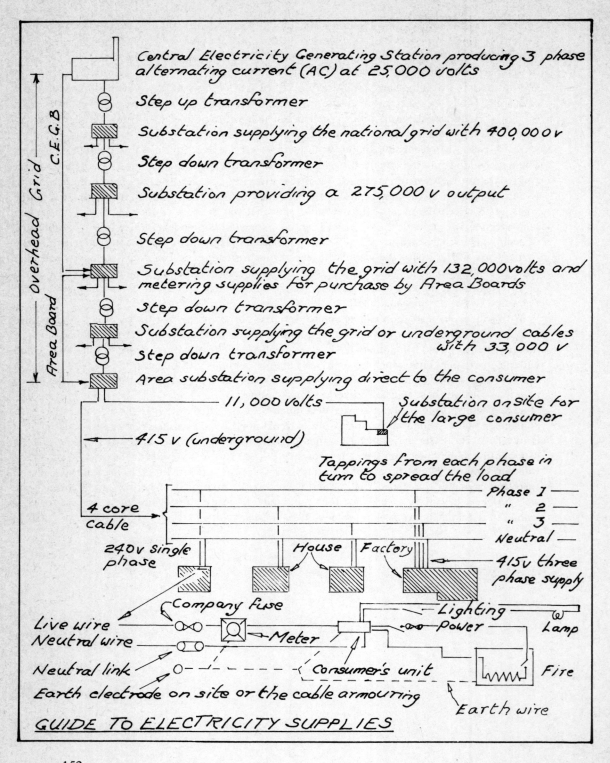

Central Electricity Generating Station producing 3 phase alternating current (AC) at 25,000 volts

Step up transformer

Substation supplying the national grid with 400,000 v

Step down transformer

Substation providing a 275,000 v output

Step down transformer

Substation supplying the grid with 132,000 volts and metering supplies for purchase by Area Boards

Step down transformer

Substation supplying the grid or underground cables with 33,000 v

Step down transformer

Area substation supplying direct to the consumer

11,000 volts

Substation on site for the large consumer

415 v (underground)

Tappings from each phase in turn to spread the load

Phase 1
" 2
" 3
Neutral

4 core cable

240v single phase

House   Factory

415v three phase supply

Company fuse

Lighting
Power

Lamp

Live wire
Neutral wire

Meter

Neutral link

Consumer's unit

Fire

Earth electrode on site or the cable armouring

Earth wire

C.E.G.B
Overhead Grid
Area Board

GUIDE TO ELECTRICITY SUPPLIES

8 Electricity

Electricity generated by the Central Electricity Generating Board's power stations is passed out on to the national grid and then sold to the Area Boards who in turn sell it to the consumers. Usually the supply from the Area Board comes underground from the substations, but not always, the rating being 240 V for domestic use otherwise 415 V.

Tappings made from cables in the road are usually two wire to domestic properties and small non-industrial establishments, otherwise four wire cables, or in rural areas tappings may be made from overhead lines and taken through step down transformers sited on poles before connecting then to the premises.

Each main cable will have a sealing chamber, main fuse and isolating link and one or more meters attached to a meter board which is fixed to but clear of the wall surface. Leads are taken from the meter to a consumer's unit which comprises a main switch for cutting off the supply if needed and fuses to control any excess currents which may be due to faults and short circuiting occurring. See figure 194.

Radial circuits run from the consumer's unit to supply the lighting, cooker and immersion heater, and ring mains for the power outlets, each ring serving 100 m² of floor area only. See figure 195.

Anticipative problems	Considerations
1 **Metering and the isolation of supplies**	Provide space in a dry area for a meter board, and electrical intake cable. Provide a meter or meters depending on tariff to be used, for example night storage heaters use electricity at off-peak rates whereas lighting will be charged at peak rate prices. Provide a consumer's unit with a main switch and fuses or circuit breakers. See figure 194. *Note* The meters may be internal or readable externally and these together with the consumer's unit must be easily accessible at all times.
2 **Choice of wiring system**	(a) **Surface wiring** Usually this system is used for temporary work, for extra circuits where no room is left in the conduits for more wires, and for areas screened from view such as roof spaces. (b) **Buried cables** This is not advisable unless the cables are protected by metal or plastic channelling buried within the plaster finishes. (c) **Wiring in conduit** Metal or plastic conduit of circular or oval section set in small chases in the wall is the most expensive but is best, in that the plaster finishes are not affected and it permits rewiring to be carried out when needed. (d) **Mineral insulated copper cable (MICC)** This is very expensive but has a very long life. It is not subject to heat and condensation problems as are other systems but does need protection from mechanical damage and damp whilst being installed.

153

Anticipative problems	Considerations
3 **Mechanical damage**	**1** Provide conduit over the cables to ensure some protection from impact when exposed and at low level, and possible damage by nails when the service is hidden and fixings to walls occur later. *Note* Metal conduit should be earthed and all lengths should be screw jointed and galvanised to extend its life. Inspection covers are advisable on bends and tees, etc, to make the job of pulling cables through and altering the direction of runs easier, especially when rewiring is required. They should be accessible. See figure 196. **2** MICC cables are not run in conduit but protection is required against impact damage where it can be reasonably anticipated. **3** Ensure that all cables run in voids such as floor spaces are at least 50 mm clear of the floor or ceiling in order to avoid possible damage from fixing nails. Joists should be drilled through the neutral axis for cables running across them and never notched. See figure 197.
4 **Condensation**	Ensure that in positions where condensation may occur in conduit, for example in roof spaces over kitchens and bathrooms, that it is protected with thermal insulation, or in the case of higher non-domestic buildings that the services to the top-most storey are in MICC unless protected.
5 **Segregation of electrical services**	Ensure that contact cannot be made between power cables (240 v), telephone cables (50 v) and wires for alarm systems (12 or 24 v) by the use of special trunking or approved spacers. See figure 198. *Note* It is dangerous for wires carrying different voltages to be mixed as it is possible for persons using low voltage systems, for example a door bell, to receive a 240 v charge should a fault occur. To provide insulation for all cables equal to that required for the highest voltage is acceptable but quite impracticable.
6 **Excess current control**	Ensure that each circuit is fused and that the correct rating for the anticipated load is used. The fuse is the weak link in the circuit and when the current becomes excessive resulting from a fault the fuse should melt and terminate the supply. *Note* If a 13 amp fuse is fitted to a plug which is attached to a 3 amp flexible lead for a table lamp and a fault developed the lighting flex may heat up and start a fire before the fuse failed.
7 **Earthing systems**	Ensure that all circuits, metal conduits and fittings and all metal appliances unless doubly insulated, are connected to an earth electrode or the armouring on the incoming supply cable. Although a water main may be earthed it is not permitted to use this as the sole earth any more. *Note* When faults occur and the appliance is made live due possibly to strands of live wire touching the metal frame then any person touching that unit will provide in themselves a path along which the fault current can flow resulting in them receiving a shock. An earth wire should

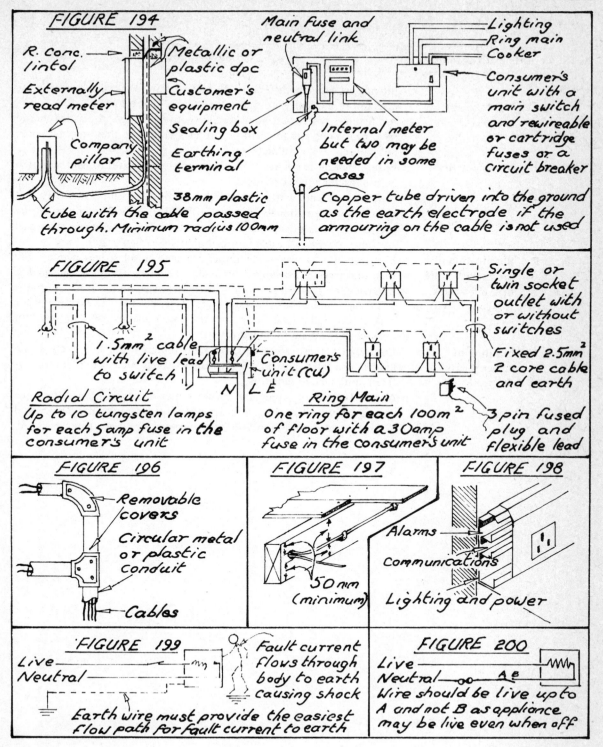

FIGURE 194

R. Conc. lintol
Externally read meter
Company pillar
Metallic or plastic dpc
Customer's equipment
Sealing box
Earthing terminal
38mm plastic tube with the cable passed through. Minimum radius 100mm

Main fuse and neutral link
Internal meter but two may be needed in some cases
Copper tube driven into the ground as the earth electrode if the armouring on the cable is not used

Lighting
Ring main
Cooker
Consumer's unit with a main switch and rewireable or cartridge fuses or a circuit breaker

FIGURE 195

1.5mm² cable with live lead to switch
Consumers unit (cu)
N L E

Radial Circuit
Up to 10 tungsten lamps for each 5amp fuse in the consumer's unit

Ring Main
One ring for each 100m² of floor with a 30amp fuse in the consumer's unit

Single or twin socket outlet with or without switches
Fixed 2.5mm² 2 core cable and earth
3 pin fused plug and flexible lead

FIGURE 196

Removable covers
Circular metal or plastic conduit
Cables

FIGURE 197

50mm (minimum)

FIGURE 198

Alarms
Communications
Lighting and power

FIGURE 199

Live
Neutral

Fault current flows through body to earth causing shock
Earth wire must provide the easiest flow path for fault current to earth

FIGURE 200

Live
Neutral
A   B
Wire should be live up to A and not B as appliance may be live even when off

Anticipative problems	Considerations
	provide an easier path for this current to flow along thus removing the discomfort and danger to the consumer. See figure 199.
	All fittings, switches, ceiling roses, etc, even though in plastics should have provisions for an earth in case at a later date they may be taken out and replaced with metal ones.
8 Wrong polarity	Ensure that all live wires from the supply first pass through the switch before reaching the appliance.
	Note Units will work if wired the wrong way round but they will remain live even though switched off. This may not be known to the user and there is a danger of shocks or burns should he happen to touch any ends of wires, or cable terminal points. See figure 200.
9 Illumination of rooms with more than one access	Ensure that two way switching is provided in all spaces where more than one access is available. This avoids the necessity and inconvenience of having to cross a dark room in order to switch on the light, or after it has been switched off. Provide lights where needed, for example over a sink in order that the body does not cast a shadow over the working area when the main room light is on.
10 Location of power points	Ensure that these are accessible when required and not likely to be hidden at a later date due to moveable fittings such as refrigerators, or fixed units such as radiators.
	Note The use of a plug and socket outlet fixed on the wall face can considerably reduce the space available, for example in a recess, and make that space too narrow for the purpose for which it was designed.

Index